Root Cause Analysis

Root Cause Analysis

A Step-By-Step Guide to Using the Right Tool at the Right Time

Matthew A. Barsalou

CRC Press is an imprint of the
Taylor & Francis Group, an **informa** business
A PRODUCTIVITY PRESS BOOK

CRC Press
Taylor & Francis Group
6000 Broken Sound Parkway NW, Suite 300
Boca Raton, FL 33487-2742

Printed on acid-free paper
Version Date: 20140624

International Standard Book Number-13: 978-1-4822-5879-0 (Paperback)

Library of Congress Cataloging-in-Publication Data

Barsalou, Matthew A., 1975-
Root cause analysis : a step-by-step guide to using the right tool at the right time / Matthew A. Barsalou.
pages cm
Includes bibliographical references and index.
ISBN 978-1-4822-5879-0 (paperback)
1. Root cause analysis. I. Title.

TA169.55.R66B37 2015
658.4'013--dc23 2014024634

Visit the Taylor & Francis Web site at
http://www.taylorandfrancis.com

and the CRC Press Web site at
http://www.crcpress.com

To my wife Elsa for her patience while I wrote this book

and my son Leander who was born as I was starting it.

Contents

Preface

There are many books and articles on Root Cause Analysis (RCA); however, most concentrate on team actions such as brainstorming and using quality tools to discuss the failure under investigation. These may be necessary steps during an RCA, but authors often fail to mention the most important member of an RCA team, the failed part. Actually looking at the failed part is a critical part of an RCA, but is seldom mentioned in RCA related literature. The purpose of this book is to provide a guide to empirically investigating quality failures using the scientific method in the form of cycles of Plan-Do-Check-Act (PDCA) supported by the use of quality tools.

This book on RCA contains two sections. The first describes the theoretical background behind using the scientific management and quality tools for RCA. The first chapter introduces the scientific method and explains its relevance to RCA when combined with PDCA. The next chapter describes the classic seven quality tools and this is followed by a chapter on the seven management tools. Chapter 4 describes other useful quality tools and Chapter 5 explains how Exploratory Data Analysis (EDA) can be used during an RCA. There is also a chapter describing how to handle RCAs resulting from customer quality complaints using an 8D report and a chapter presenting an example of an RCA.

The second section contains step-by-step instructions for applying the principles described in the first section. The tools presented are briefly described, key points are summarized and an example is given to illustrate the concept. This is followed by the procedure for applying the concept. The intent is to make the step-by-step procedure available for less experienced investigators, with the examples being sufficiently clear for more experienced investigators to use as a quick reference.

There are many tools available for root cause analysis and unfortunately it would be impractical to describe an appropriate phase in an investigation to use each tool. Therefore this book offers an appendix with a guide to tool selection based upon the intended use of the tool. There is also an appendix explaining the terminology used in the book.

Introduction

This guide to root cause analysis (RCA) is intended to provide root cause investigators with a tool kit for the quick and accurate selection of the appropriate tool during a root cause investigation. The handbook consists of two parts. Part 1 provides more detailed information regarding the tools and methods presented here. Part 2 contains less background information but has step-by-step instructions and is intended for use as a quick reference when a tool is needed.

Root cause analysis is the search for the underlying cause of a quality problem. There is no single RCA method for all situations; however, the RCA should involve empirical methods and the selection of the appropriate tools for the problem under investigation. For example, a run chart may be appropriate for investigating a length deviation that sporadically occurs over a longer period of time. A run chart would not be useful if the length deviation is on a unique part. An RCA is performed by a root cause investigator; in manufacturing, this could be a quality engineer, quality manager, or even a well-trained production operator.

For every problem, there is a cause, and RCA tools are used to identify the cause (Doggett, 2005). Most quality problems can be solved using the classic seven quality tools (Borrer, 2009). Typical quality problems include a machined part out of specification or a bracket with insufficient tensile strength. The use of quality tools alone to solve a problem may not be sufficient; Bisgaard (1997) recommends the addition of the scientific method when performing RCA. The use of the seven quality tools and the scientific method may not be sufficient for investigating a complex quality problem. ReVelle (2012) warns that "quality professionals with only a limited number of tools at their disposal perform the task in a suboptimal way, giving themselves—and the tools they've used—a bad name" (p. 49). The scientific method together with the classic seven quality tools and other tools and methods form a complete quality toolbox.

The scientific method is fundamental to RCA yet is often overlooked by industry. Repeatedly making random adjustments to the system is often used in place of careful observation and controlled experiments (Hare, 2002). Small random adjustments to a system may be a useful method of gleaning information about a system; however, the adjustments should be systematic and not random, as in without purpose. That is, the results of each change to the process should be recorded, and the variables affected should be written down so comparisons can be made and conclusions can be drawn. Variables should be controlled to ensure that the changes to the system are not the result of an unknown factor and misattributed to the factor under consideration.

An example of an improper experiment is a hypothetical manufacturing company with a rust problem on the steel tubes the company produces. A quality engineer decides to study the effects of rust on tube diameter by placing samples of various tube sizes in a humidity chamber for a week to simulate aging. The experiment is set to run for six months. Every week the diameter of five randomly selected tubes of the same diameter is recorded, the tubes are wiped down to remove metal chips and are then sprayed with rust inhibitor and placed in the humidity chamber. The samples are checked daily for rust, and the first day a part is observed to have rust is recorded so that at the end of the study the quality engineer can determine if a correlation exists between rust and diameter. At the end of the week, a new set of five tubes of a different diameter is randomly selected, and the process is repeated. It is unknown if there is a correlation between rust and tube diameter, and if there is a correlation, it is expected to be a weak correlation; the negative correlation between rust and rust inhibitor is a known strong correlation. Unfortunately, the rust inhibitor bottle was refilled from a large container in production that was refilled once a week and the ratio of rust inhibitor to water changed daily as the water evaporated from the container. The effect of tube diameter on rust was lost in the data because of the stranger, yet uncontrolled, effect of the variations in the strength of the rust inhibitor.

The proper use of the scientific method could have prevented wasting the time spent on an experiment that could have yielded misleading results. The scientific method is a powerful method for understanding data; however, it is not the only methodology available to industry and the field of quality. The classic seven quality tools are known by many names, such as "the old seven," "the first seven," or "the basic seven" (Rooney, 2009), and are important enough as problem-solving tools in the field of quality that the American Society for Quality considers them to be part of the quality engineer's body of knowledge (Borrer, 2009). These tools are useful for the collection and analysis of data and should be used to support the scientific method.

There are many tools of varying degrees of usefulness in addition to the classic seven quality tools. Additional quality tools such as 5 Why, cross assembling, is-is not analysis, and matrix diagrams are presented in this book. In addition, an explanation is given regarding how a root cause investigator can successfully follow multiple lines of evidence to arrive at a root cause.

The quality tools can be supported by exploratory data analysis (EDA). Used together, the scientific method, quality tools, and statistics in the form of EDA can build a picture of the actual situation under investigation and lead the root cause investigator to the true root cause of an event.

About the Author

Matthew Barsalou is employed by BorgWarner Turbo Systems Engineering GmbH, where he provides engineering teams with support and training in quality-related tools and methods, including root cause analysis. He has a master of liberal studies from Fort Hays State University and a master of science in business administration and engineering from Wilhelm Büchner Hochschule. His past positions include quality/laboratory technician, quality engineer, and quality manager.

Matthew Barsalou's certifications include TÜV quality management representative, quality manager, quality auditor, and ISO/TS (International Organization for Standardization/Technical Specification) 16949 quality auditor as well as American Society for Quality certifications as quality technician, quality engineer, and Six Sigma Black Belt.

He is the American Society for Quality's *Statistics Division Newsletter* editor and a frequent contributor to *Quality Digest*, the Minitab Blog, and has published in German, American, and British quality journals.

Section I

Introduction to Root Cause Analysis

1

The Scientific Method and Root Cause Analysis

The textbook explanation of the scientific method is "Collect the facts or data. … Formulate a hypothesis. … Plan and do additional experiments to test the hypothesis. … Modify the hypothesis." And, an experimenter should consider a hypothesis to be a "tentative explanation of certain facts" and must remember that a "well-established hypothesis is called a theory or a model" (Hein and Arena, 2000). This textbook example of the scientific method is correct, but too oversimplified for use in root cause analysis (RCA).

Before going into detail regarding the scientific method, there are terms used in the scientific method that may require clarification so that their exact meanings are easily understood. One important term is *hypothesis*. A hypothesis "is a postulated principle or state of affairs which is assumed in order to account for a collection of facts" (Tramel, 2006, p. 21). The word *theory* is often used colloquially to mean hypothesis; however, hypothesis should not be used interchangeably with the word *theory*; a hypothesis is preliminary and more tentative than a theory.

A hypothesis is used to both explain the past and predict the future, and it should generally meet five virtues: conservatism, modesty, simplicity, generality, and refutability. Conservatism means that it is better when a hypothesis has few conflicts with previous beliefs and hypotheses. Modesty means that the better hypothesis is the hypothesis that makes the less-complicated assumptions. Simplicity is much like modesty: A hypothesis is more likely to be correct when it makes simpler assumptions. Generality means that a hypothesis should avoid being too specific. For a hypothesis to serve any purpose, it must be refutable; there must be some way in which to disprove it (Quine and Ullian, 1978).

The five virtues of a hypothesis are no guarantee that a hypothesis will be correct; however, conformity to the virtues increases the chance that a hypothesis will be correct and can help in choosing between two hypotheses. There is a greater chance of a hypothesis being incorrect if it requires complicated lines of reasoning and makes many complex assumptions. The first virtues are along the same lines as Occam's razor, which is a principle that "urges us when faced with two hypotheses that explain that the data equally well to choose the simpler" (Sagan, 1996, p. 211) hypothesis.

The final virtue is much like Popper's falsification: A hypothesis must be falsifiable or it is not scientific; it is not possible to truly prove something because contrary evidence could always be discovered at a later time. An experimenter should not consider the results of an experiment as conclusive evidence but rather as tentative. However, the more robust the testing is, the higher the degree of corroboration will be (Popper, 2007). An experimenter who only performs one simple experiment has a lower degree of corroboration than an experimenter who rigorously tests his or her hypothesis. For example, a hypothetical experimenter suspects that tube diameter correlates with rust on a steel tube. The experimenter places a 20-mm diameter tube and an 80-mm diameter tube in an environmental chamber and checks for rust every day until the fourth day of the experiment, when the smaller tube is found to be rusty. Such an experiment would result in a low degree of corroboration. A more robust experiment would involve controlling the variables, random sampling for sample collection, more samples during each experimental run, and repeated test runs. This would result in a higher degree of collaboration, and the experimenter can be more confident of the accuracy of the results.

Two other important terms are *dependent* and *independent variable*. The dependent variable is the result or outcome of the experiment, and the independent variable is an "aspect of an experimental situation manipulated or varied by the researcher" (Wade and Travis, 1996, p. 60). The dependent variable is also known as the response variable. The independent variable in the rust study experiment is the tube diameter; the experimenter varied the tube diameter to see what affect it had on the final outcome, which was the dependent variable. The independent variables "are also called the treatment, manipulated antecedent, or predictor variables" (Creswell, 2003, p. 94).

Treatments, also known as experimental runs, are the specific sets of conditions during an experiment. For example, the treatment variable in the rust study is rust on the tubes, and the experimental run is the experiment in which a sample tube is wiped down, coated in rust inhibitor, and placed into a humidity chamber.

A *factor* is "a process condition that significantly affects or controls the process output, such as temperature, pressure, type of raw material, concentration of active ingredient" (Del Vecchio, 1997, pp. 158–159). The treatment variable is one factor; another type of factor is the confounding variable. The confounding variable "is not actually measured or observed in a study," and "its influence cannot be directly detected in a study" (Creswell, 2003) Uncontrollable factors such as the confounding variable are referred to as *noise* (Montgomery, 1997) The effects of the noise being mixed into the results is referred to as *confounding* (Gryna, 2001). The effects of confounding can be lessened by the use of blocking, which is "a technique used to increase the precision of any experiment" (Montgomery, 1997, p. 13) by mixing the confounding factors across all experimental sets to ensure that although the effects cannot be eliminated, they will have less impact by being more evenly spread across the experimental results.

An example of noise in the rust study was the rust inhibitor; the rust inhibitor had a great deal of influence on the formation of rust but was not applied in consistent quantities because the rust inhibitor used was collected at different times and had varying degrees of rust inhibitor concentration in the mixture. Blocking is an option that could be used if the concentration in the rust inhibitor mixture could not be controlled. The experimenter could have used sample tubes with different diameters instead of using sample tubes of the same diameter each week. This way, the effect of the variation in the rust inhibitor would have been spread across the samples and stronger or weaker concentrations of rust inhibitor would equally affect all tube diameter sizes.

Randomization and replication should also be used when performing an experiment. Randomization is also used to cancel out the effects of noise, and it can increase the reliability of the experimental results. Replication is the repetition of an experiment to increase the precision of the results (Gryna, 2001). The experimenter conducting the rust study should have randomly selected the sample tubes. For example, instead of selecting them from the box nearest the laboratory, samples should have been taken from different randomly selected areas. It is possible that the parts stored nearest the laboratory were near a door and exposed to moisture so that they were already starting to rust just prior to the start of the experiment.

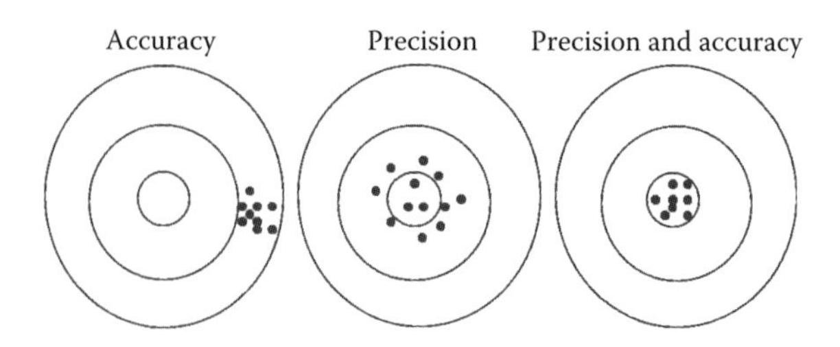

FIGURE 1.1
Precision and accuracy.

Such random factors could be canceled out by randomization. Replication of the experiment would have shown the variation in the results; to be accurate and precise, the results should be consistent with each other.

The precision of measurement results is the closeness of each result to each other. Also related to precision is accuracy, which is the closeness of the measurement results to the true value of what is being measured (Griffith, 2003). Juran and Gryna (1980) illustrate the difference between accuracy and precision using targets. The hits clustered closely together but away from the center are precise but not accurate. The hits that are scattered around the center of the target are accurate but not precise. Only the hits clustered close together and at the center of the target are both accurate and precise. Figure 1.1 graphically depicts the difference between precision and accuracy.

Ideally, the measurements taken after an experiment will be both precise and accurate. The experimenters in the rust study did not properly control their variables. This problem could have been avoided by ensuring that a consistent amount of rust inhibitor was in the mixture used or blocking was used to lessen the effects of the variation on the precision and accuracy of the results.

To fully use the scientific method, an understanding of deduction and induction is necessary. Deduction is a "process of reasoning in which a conclusion is drawn from a set of premises … usually … in cases in which the conclusion is supposed to follow from the premise," and induction is "any process of reasoning that takes us from empirical premises to empirical conclusions supported by the premises" (Blackburn, 2005, p. 90). Deduction "goes from the general to … the particular" and induction goes "from the particular to the general" (Russell, 1999, p. 55). Deduction uses logical connections, such as, "A red warning light on a machine indicates a problem; therefore, a machine with a red warning light has a problem."

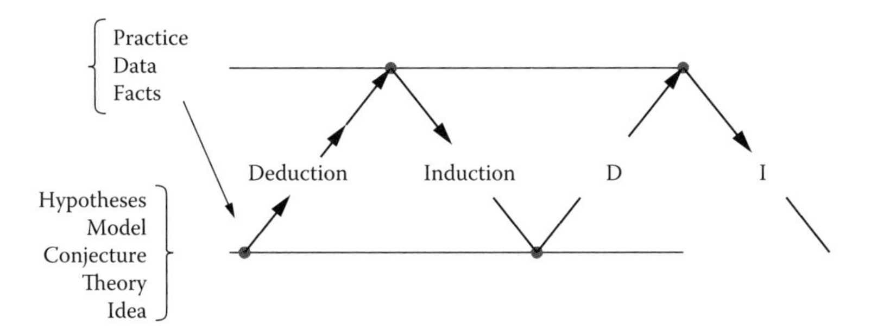

FIGURE 1.2
The iterative inductive-deductive process. (From George E.P. Box. *Journal of the American Statistical Association* 71 no. 356 (1976): 791. With permission of Taylor & Francis.)

Deduction moves from generalities to specifics and uses what is known to reach conclusions or form hypotheses. Induction goes from specifics to generalities and uses observations to form a tentative hypothesis. Induction uses observations such as, "Every machine with a red light has a problem; therefore, the machine I observed with a red light must have a problem."

Box, Hunter, and Hunter (2005) explain that deduction is used to move from the first hypothesis to results that can be compared against data; if the data and expected results do not agree, induction is used to modify the hypothesis. This "iterative inductive-deductive process" (Box, Hunter, and Hunter, 2005) is iterative; that is, it is repeated in cycles as shown in Figure 1.2.

The iterative inductive-deductive process is used together with experimentation during RCA. Each iteration or repetition of the inductive-deductive process should provide new data that can be used to refine the hypothesis or to create a new hypothesis that fits the new data. Each iteration should bring the root cause investigator closer to the root cause, even if a hypothesis must be completely discarded. The investigator can use the negative information to tentatively exclude possible root causes. Using the iterative inductive-deductive process with experimentation can provide a glimpse of the "true state of nature" (Box, 1976) (Figure 1.3).

There are many elements that are essential for the proper use of the scientific method. For example, controls are needed to ensure that the resulting response variable is the result of the influence of the treatment variable (Valiela, 2009). Controlling the variables helps to separate the influences of noise from the influence of the treatment variable on the response

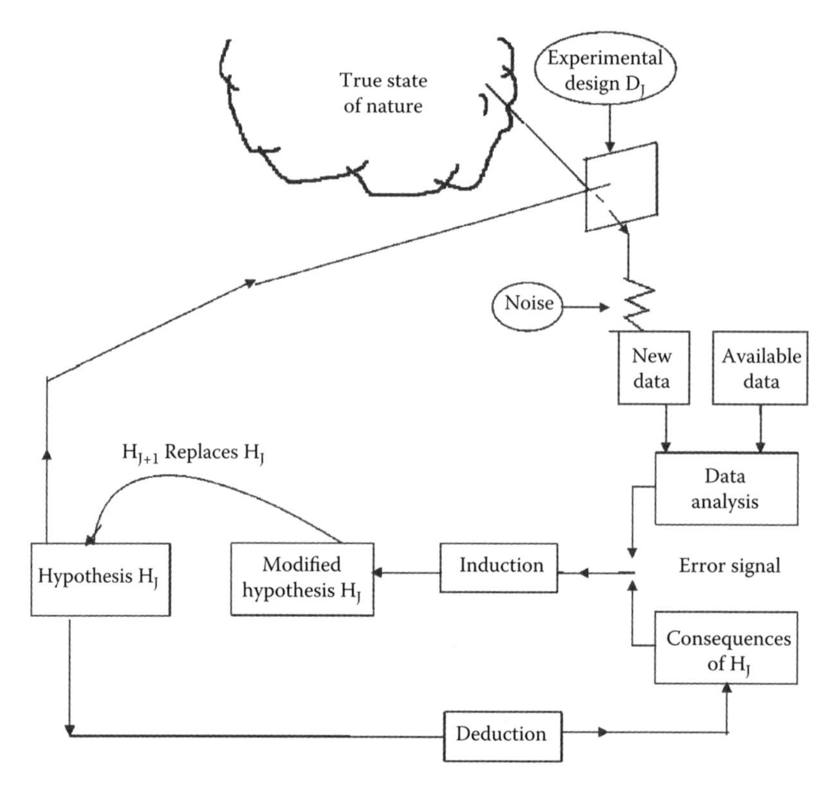

FIGURE 1.3
Experimentation and the true state of nature. (From George E.P. Box. *Journal of the American Statistical Association* 71 no. 356 (1976): 796. With permission of Taylor & Francis.)

variable. In the rust study example, the treatment variable rust inhibitor was not properly controlled; various concentrations of rust inhibitor were unknowingly used, and the results of the study were masked by the variation in the rust inhibitor.

Feynman (1988) points out the need for objectivity in using the scientific method. He warns experimenters to be careful of preferring one result more than another and uses the example of dirt falling into an experiment to illustrate this hazard; the experimenter should not just accept the results that he or she likes and ignore the others. The favorite results may just be the result of the dirt falling into the experiment and not the true results of the experiment. In such a situation, the results without the contaminating dirt were discarded because they were not what the experimenter wanted to see. Objectivity is essential for achieving experimental results that truly reflect reality and not the wishes of the experimenter.

To help achieve objectivity, an experimenter can use blinding so that the experimenter does not know which results correspond to which variables (Wade and Travis, 1996). An experimenter may not consciously be biased but may still show an unconscious tendency to favor results that confirm what was already suspected. A simple method of blinding is to evaluate the results of an experiment without knowing if the results are for the control or the factor under consideration. An experimenter can also use blinding when checking measurement data by having the checks performed by a second person who does not know what to expect.

Also needed in using the scientific method are operational definitions. Operational definitions are clear descriptions of the terms that are used (Valiela, 2009). Doctor W. Edwards Deming, who has been referred to as the "Man Who Discovered Quality" (Gabor, 1990), explains that operational definitions are needed because vague words such as bad or square fail to express sufficient meaning; operational definitions should be such that all people who use them can understand clearly what is intended. Deming uses the example of a requirement for a blanket to be made of 50% wool. Without an operational definition, one-half may be 100% cotton and the other half 100% wool; an operational definition would include a method of establishing the exact meaning of 50% wool. Deming's (1989) operational definition for 50% wool is to specify the number of uniform test samples to be cut from a blanket and the exact method of testing to determine the amount of wool in the samples of the blanket. An operational definition for the rust inhibitor used in the rust study would be, "The rust inhibitor solution must contain between 66% and 68% rust inhibitor, with the remainder of the solution being water." Operational definitions bring clarity to what is being described and help to eliminate confusion.

The scientific method is empirical; that is, it is based on observation and evidence. However, Tramel (2006) sees the scientific method as containing both empirical and conceptual elements. A hypothesis is formed by the empirical element, observing data. The next steps are conceptual: Analyze the data and then form a hypothesis. The following steps are conceptual elements of testing the hypothesis; assume the hypothesis is true so it can be tested and then deduce the expected results based on the hypothesis. The final element is an empirical observation in which the results of the test are compared against the expected results based on the hypothesis.

Adriaan de Groot's (1969) description of the scientific method is an empirical cycle consisting of five phases: observation, induction, deduction, testing, and evaluation. Observations should be made so that data can be collected, although de Groot acknowledges that an investigator may have a preconceived notion of what the hypothesis will be. Induction is then used to form a hypothesis based on the data at hand. Deduction is used when the hypothesis makes a prediction that would then be tested empirically and the results evaluated and the hypothesis modified or replaced with a new one if necessary.

In using the scientific method, Platt (1964) recommends forming multiple hypotheses, performing preliminary experiments to determine which hypothesis should be excluded, performing another experiment to evaluate the remaining hypothesis, and then modifying hypotheses and repeating the process if necessary. If there are insufficient data to form a plausible hypothesis, a root cause investigator could perform exploratory investigations to gather data. An exploratory investigation is less structured than the formal scientific method; however, it should not be confused with searching for data in a true haphazard way (de Groot, 1969). During an exploratory investigation, the root cause investigator forms many hypotheses and tries to reject them or hold them for more structured testing at a later time. An exploratory investigation is part of data gathering and is not the search for the final root cause.

Once sufficient data are gathered, a hypothesis should be generated and tested; the hypothesis must be capable of predicting the results of the test, or it should be rejected because, "If one knows something to be true, he is in a position to predict; where prediction is impossible, there is no knowledge" (de Groot, 1969, p. 20). The power to predict is the true test of a hypothesis. Medawar (1990) tells us a scientist "must be resolutely critical, seeking reasons to disbelieve hypotheses, perhaps especially those which he has thought of himself and thinks rather brilliant" (p. 85); the same can be said for a root cause investigator. It is acceptable to begin an RCA with a preconceived notion regarding what the root cause is; however, preconceived notions should be based on some form of data because "it is a capital mistake to theorize before one has data. Insensibly one begins to twist facts to suit theories, instead of theories to suit facts" (Doyle, 1994, p. 7). The investigator must be absolutely sure to follow the evidence and not an opinion. Blinding could be useful here to guard against the hazards of a pet hypothesis.

Box compares his iterative inductive-deductive process to Deming's Plan-Do-Check-Act (PDCA) in that they are both cycles for the generation of new knowledge (Box, 2000). Deming referred to the PDCA cycle as the

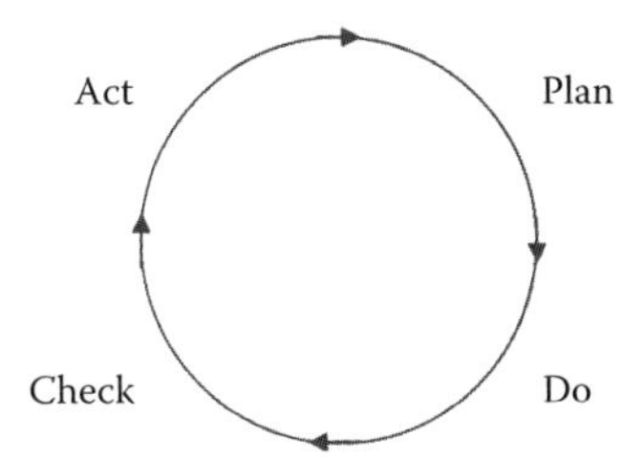

FIGURE 1.4
The Deming cycle.

Shewhart cycle because it was originally based on a concept developed by Walter A. Shewhart in the 1930s. Deming taught it to the Japanese in the 1950s, and the concept became known as the Deming cycle (Deming, 1989; Figure 1.4). The PDCA cycle was improved in Japan and began to be used for contributions to quality improvements; it is unclear who made the changes, although it may have been Karou Ishikawa (Kolesar, 1994). By the 1990s, Deming had renamed PDCA as Plan-Do-Study-Act (PDSA) (Deming, 1994) although the name PDCA still remains in use. The Deming cycle can be viewed in Figure 1.4.

The PDCA concept is known in industry; therefore, it can be used in industry without introducing a completely new concept. It can also be combined with the scientific method and the iterative inductive-deductive process to offer an RCA methodology for opening the window into the true state of nature, also known as finding the root cause. The PDCA cycle has the additional advantage that it can be used for the implementation of corrective actions and quality improvements after a root cause has been identified. *Kaizen*, also known as continuous improvement, frequently uses the PDCA cycle (Imai, 1986).

The first iterative of the PDCA cycle should consist of making observations (Figure 1.5). This could mean collecting new data or using what is already known if sufficient information is available. The observations are to be used with deduction to go from a tentative hypothesis that extrapolates beyond the data. The hypothesis is then tested. This could be as complicated as an experiment in a laboratory or as simple as observing a defective component. The complexity level should be determined by the problem under investigation. The test results are compared against the hypothesis and actions are taken based on the results. A more elaborate confirmation test may be required if the test failed to refute the tenta-

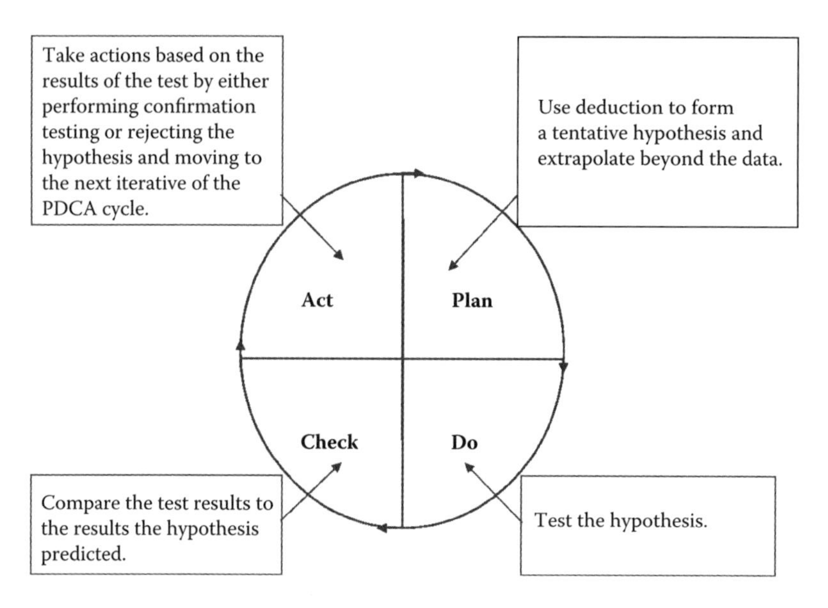

FIGURE 1.5
The first PDCA cycle.

tive hypothesis. The next iterative of the PDCA process is required if the hypothesis is rejected.

The next PDCA iterative uses induction to form a new or modified hypothesis based on the test results of the previous iterative. The hypothesis is then tested, and the test results are compared against the predicted results; this PDCA cycle is shown in Figure 1.6. Confirmation testing is performed if the hypothesis is not refuted, and the next PDCA cycle starts if the hypothesis is refuted.

Deduction is used to form a new or modified hypothesis, taking into consideration the failure of the previous hypothesis. The new or modified hypothesis is then tested and the results are compared against the predictions made by the hypothesis as displayed in Figure 1.7. Confirmation testing or a new PDCA cycle is then required.

The next PDCA cycle, as depicted in Figure 1.8, uses inductive reasoning and the results of the previous cycle to form a new or modified tentative hypothesis. The cycle then continues and is followed by new iterations of the PDCA cycle until a root cause is identified and confirmed. Confirmation is essential to ensure that resources are not wasted on implementing a corrective action that will not eliminate the problem. Identification of the wrong root cause also leads to the potential for a false sense of security because of the incorrect belief that the problem has been solved.

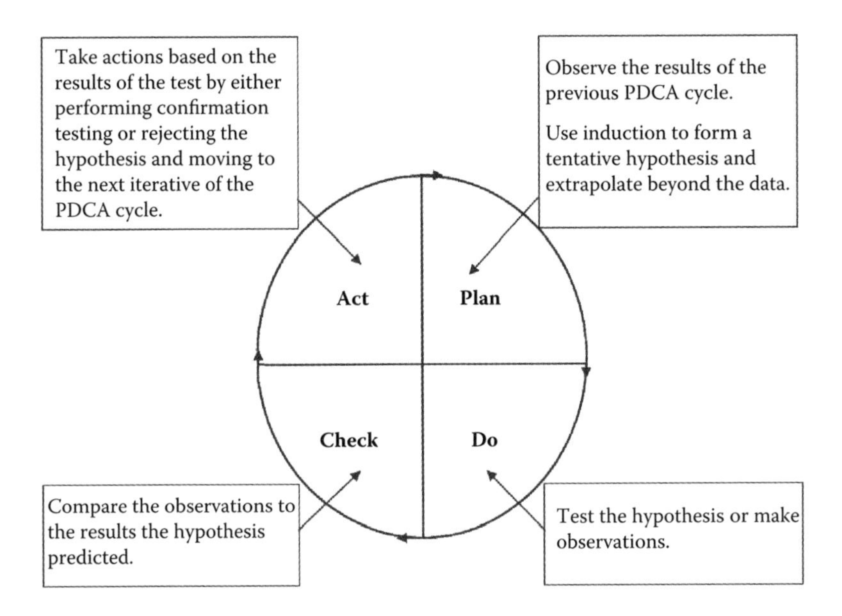

FIGURE 1.6
The second PDCA cycle.

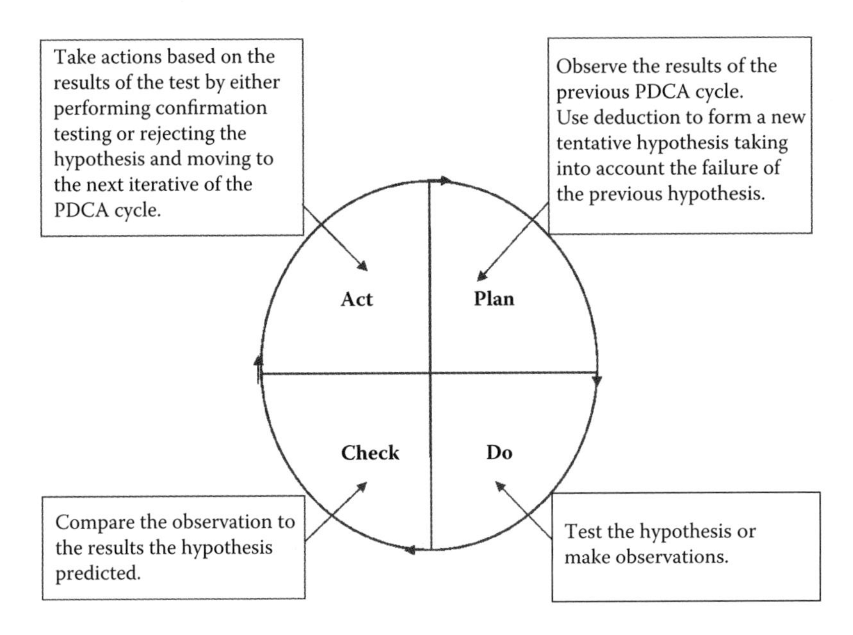

FIGURE 1.7
The third PDCA cycle.

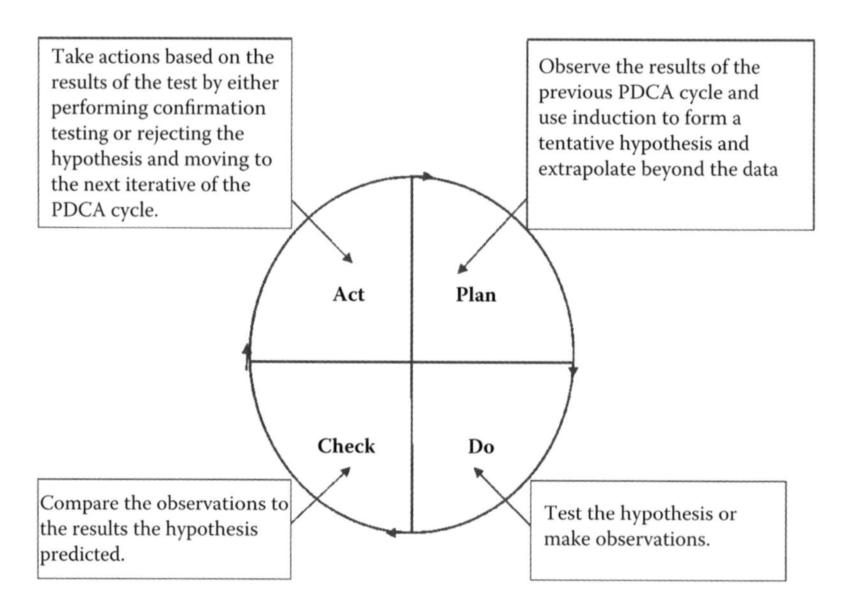

FIGURE 1.8
The fourth PDCA cycle.

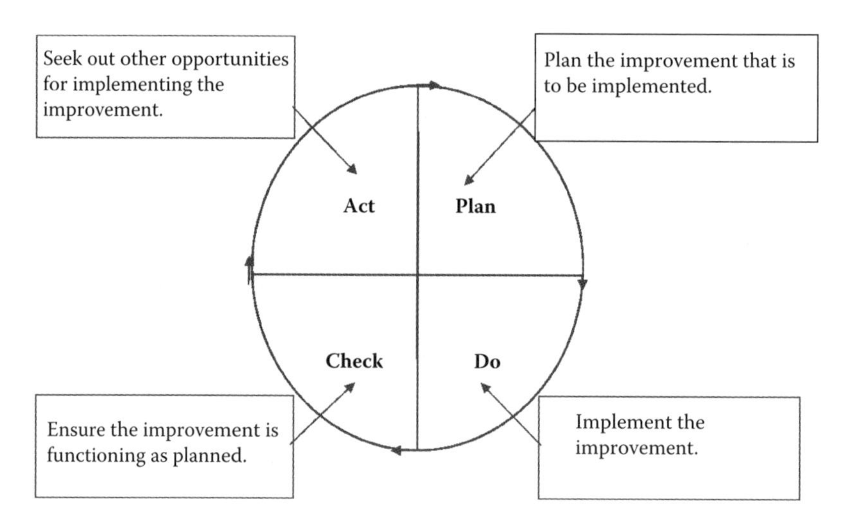

FIGURE 1.9
The PDCA improvement cycle.

Corrective and improvement actions can and should be based on PDCA, as shown in Figure 1.9. The improvement needs to be planned and then implemented. After implementation, the effectiveness of the improvement needs to be verified. An improvement that was successful during a small trial could have unintended consequences during mass production.

Opportunities should be sought for implementing a successful improvement in other locations or on other products, particularly if the same problem has the potential to occur somewhere else. The absence of a problem in other locations where it could occur may not be a sign that the problem is not there; rather, it could mean that the problem was undetected or has not occurred yet.

The iterations of the PDCA cycle should be used in such a way that the process fits the problem under investigation. For example, a sporadic problem with a continuous process would not be treated exactly the same as the RCA of an individual failed part that has been returned from a customer. For the former, a larger interdepartmental team and a formal structure would be advantageous. For the latter, a single root cause investigator may go though many quick iterations of the process while observing, measuring, and otherwise evaluating the failed part, ideally with the support of those who have the required technical or process knowledge.

Although the scientific method is a powerful, time-tested method for achieving new insights, the scientific method may not alone solve a quality problem. The scientific method provides a structured process for RCA. This process can be greatly supported by the proper use of quality tools. For example, a root cause investigator may use one quality tool to gather more information regarding the problem and a different tool to actually analyze the part in question. A root cause investigator without quality tools is like an automotive mechanic without a toolbox.

2

The Classic Seven Quality Tools for Root Cause Analysis

The classic seven quality tools were originally published as articles in the late 1960s in the Japanese quality circle magazine *Quality Control for the Foreman*. Around that time, most Japanese books on quality were too complicated for factory foremen and production workers, so Kaoru Ishikawa used the articles from *Quality Control for the Foreman* as a basis for the original 1968 version of *Guide to Quality Control*. The book was frequently used as a textbook for tools in quality (Ishikawa, 1991).

The seven tools as defined by the American Society for Quality are the flowchart for graphically depicting a process, Pareto charts for identifying the largest frequency in a set of data, Ishikawa diagrams for graphically depicting causes related to an effect, run charts for displaying occurrences over time, check sheets for totaling count data that can later be analyzed, scatter diagrams for visualizing the relationship between variables, and histograms for depicting the frequency of occurrences (Borror, 2009). These tools should be a staple in the tool kit of any root cause investigator. Although often used by quality engineers, the tools are simple enough to be effectively used by production personnel, such as machine operators.

The Ishikawa diagram was originally created by Kaoru Ishikawa, who used it to depict the causes that lead to an effect. Ishikawa originally called it a cause-and-effect diagram. Joseph Juran used the name Ishikawa diagram in his 1962 book *Quality Control Handbook*, and now the Ishikawa diagram is also known as fish-bone diagram because of its resemblance to a fish bone (Ishikawa, 1985). It is sometimes also called a cause-and-effect diagram.

An Ishikawa diagram has a horizontal line leading to the effect on the right side of the diagram. Vertical lines move at approximately a 45° angle to the left side of the diagram. These lines are typically labeled with the potential factors. Many, but not all, authors attempt to list the factors as the six *M*s: man, material, milieu, methods, machine, and measurement. Griffith (2003, p. 28) recommends using the six factors: "1. People. 2. Methods. 3. Materials. 4. Machines. 5. Measurements. 6. Environment." People can be an influence factor because of their level of training, failure to adhere to work instructions, or fatigue. Methods such as the way in which a part is assembled may be inappropriate or a work instruction may not be clear enough. Materials can be an influence for many reasons, such as being the wrong dimension or having some other defect. Machines may be worn or not properly maintained. Measurement errors could result from either a problem with the measuring device or improper use by an operator. The environment may be an influence because of the distractions of extreme temperature or loud noise (Griffith, 2003). Each of the angled lines has a horizontal line labeled with the influences on each factor. For example, measurement could be affected by the calibration of the measuring device and the accuracy of the device.

Borror (2009) demonstrates an Ishikawa diagram using safety discrepancies on a bus as the effect; the causes are materials, methods, people, environment, and equipment. Borror lists road surfaces and driving conditions as two causes under environment. Road surfaces have two subcategories: dirt and pavement. An Ishikawa diagram is provided in Figure 2.1.

An Ishikawa diagram can be subjective and limited by the knowledge and experience of the quality engineer who is creating it. This weakness can be compensated for by the use of a team when creating an Ishikawa diagram. The Ishikawa diagram may not lead directly to a root cause; however, it could be effective in identifying potential factors for further investigation.

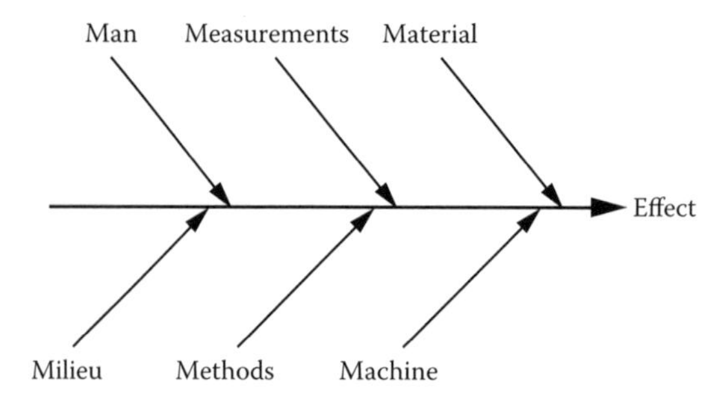

FIGURE 2.1
Ishikawa diagram.

Another of Ishikawa's tools is the check sheet. A check sheet can be used for multiple reasons, for example, to collect data on types of defects, defect locations, causes of defects, or other uses as deemed appropriate (Ishikawa, 1991). Griffith (2003) uses failure types for an example of a check sheet. The example uses defect, count, and total as the column headings, and the types of defects are listed under the defect heading. In the count category, tally marks are used to indicate the number of occurrences of each defect. The total number of each defect type is then entered in the total column. Check sheets can be used for data collection.

For example, the data collected by a check sheet could be used for the prioritization of a root cause investigation; it may be better to seek the root cause of an occurrence that happened 25 times before investigating the root cause of an event with only 1 occurrence. Naturally, common sense must be applied so that a high occurrence minor failure is not investigated instead of a one-time occurrence that could put people in danger, such as a faulty brake system in an automobile or a faulty shutoff switch on an industrial press. A check sheet used for types of failures is shown in Figure 2.2.

Check sheets can also be used for collecting data on the location of a failure. Ishikawa (1991) demonstrates the use of a check sheet for a defect location using a problem with bubbles in the laminated glass that is to be used for automotive windshields. The problem location was narrowed down by identifying the location of the defects on a schematic representation of a windshield. Using this information, the root cause investigator was able to localize the problem and found the root cause to be a lack of pressure during lamination in the affected area. Such a graphic check sheet used for scratches on a side panel is shown in Figure 2.3. In this example, it is obvious the problem is only occurring in one area of the side panel.

Failure	Number of Occurrences	Total
Part missing	~~////~~ ~~////~~ //	12
Scratches	~~////~~ ~~////~~ ~~////~~ ~~////~~ /	21
Dimensional problem	~~////~~ ~~////~~	10
Wrong part	~~////~~ ~~////~~ /	11
Improper assembly	~~////~~ ///	8
Assembly problems	~~////~~ /	6
Total		68

FIGURE 2.2
Check sheet.

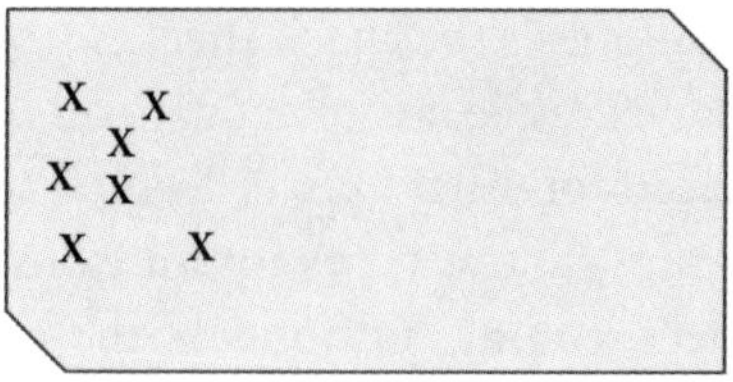

FIGURE 2.3
Check sheet for location.

More than one of the seven classic quality tools can be used to assess the same data. For example, data from check sheets can be analyzed using run charts. To do this, however, the time of data collection must be recorded in the check sheet. Run charts are much like statistical process control (SPC) but without the calculated control limits used in SPC.

Run charts are used for monitoring process performance "over time to detect trends, shifts, or cycles" as well as allowing "a team to compare a performance measure before and after implementation of a solution" (Sheehy et al., 2002). To construct a run chart, an *x* axis and *y* axis must be drawn. The *x* axis is used to indicate time, which increases from left to right. The units can be defined as needed; for example, the units could be hours, days, or production shifts. The *y* axis is where the measurement results are placed; these can be actual measurements arranged from lowest to highest or the number of occurrences. A run chart is shown in Figure 2.4.

A quality manager may use a check sheet to collect data on defective parts being produced by several production machines. After two weeks of production, a run chart could be used to view the results. The run chart would be drawn and the data plotted. The quality manager should then observe

FIGURE 2.4
Run chart.

the data and look for patterns or trends. It may also be advantageous to use "stratification" (Stockhoff, 1988, p. 565). Stratification involves separating the data; for example, instead of plotting the total number of defective parts per shift for all three production machines, a quality manager plots the individual results of the machines separately. By using stratification, the quality manager may notice that the majority of the defective parts were produced by only one machine.

Another method for displaying and analyzing data is the histogram. A histogram is used to graphically depict frequency distributions; a frequency distribution is the ordered rate of occurrence of a value in a data set. Histograms depict the shape of a data set's distribution and as such can be used to compare the results of two different processes (Tague, 2005). To construct a histogram, an *x* axis and *y* axis are used. The *x* axis is the unit under consideration, such as defects, days between failures, or individual measurements or occurrences, such as production stops. The *y* axis is the number of occurrences, as shown in Figure 2.5.

The shape of the histogram should be observed. A statistically normal distribution should present a bell-shaped curve that rises and then peaks in the middle before descending; the results should be evenly spread about the average. A skewed histogram has a peak that is on one side or the other; a histogram would be skewed if it depicted the number of days for a repair job and most repair jobs only took a few days with a minority of repair jobs taking up to nine days.

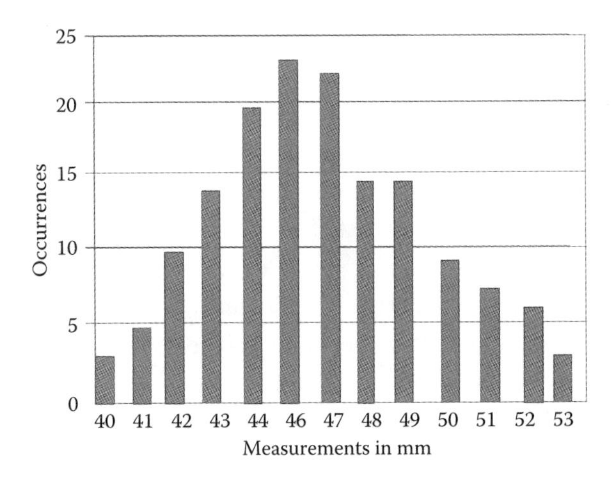

FIGURE 2.5
Histogram.

Stratification should also be considered when using histograms. Ishikawa (1991) recounts a situation in which a histogram created to show the hardness of sheet metal had a normal distribution. Sheet metal supplied to the company by a parent company was often wrinkled and cracked. The company created two histograms on discovering that the sheet metal originally came from two different subsuppliers. One subsupplier supplied parts that where skewed to the left, and the other supplier supplied parts that where skewed to the right; the true distribution was masked by the mixing of the parts.

The same data that was used in a run chart or histogram can be used in a Pareto chart. A Pareto chart is used to prioritize by separating "contributing effects into the 'vital few' and the 'useful many'" (Stockhoff, 1988, p. 565). The Pareto chart is based on the Pareto principle: 80% of effects are the results of 20% of causes. This is not a rule that holds true 100% of the time; however, it holds true often enough to be useful and should be considered.

The Pareto principle was created and misnamed by Joseph Juran, who first observed during the 1920s that only a few defects accounted for most defects in a list. Vilfredo Pareto had originally noticed that there was an uneven distribution of wealth, with 80% of all wealth in Italy owned by only 20% of the people. Juran named his concept the Pareto principle, but later believed it had been misnamed because Vilfredo Pareto was only referring to the distribution of wealth. Juran also came up with the concept of the vital few and trivial many to label the 20% of causes and 80% of causes, respectively (Juran, 2005). Juran later realized that the trivial many may also be important; therefore, he renamed the trivial many the "useful many" (Sandholm, 2005, p. 72).

A Pareto chart can be used for many things, for example, defect by types, failures by machine, customer complaints by customer, or failures by costs. The factors under consideration are listed on a horizontal line, starting from the most to the least. The vertical line on the left side is used to indicate the number of occurrences or units of each factor. The horizontal line on the right side is used to indicate cumulative percentage. The percentage for each factor is calculated and then added in a cumulative line running across the chart from left to right.

Juran (1995) illustrates the Pareto principle using a Pareto chart for sales of products within a company (Figure 2.6). The products are organized in a Pareto chart by sales volume, and the Pareto chart shows that of 20 product lines, 4 products account for 75% of sales, and the bottom 16 products

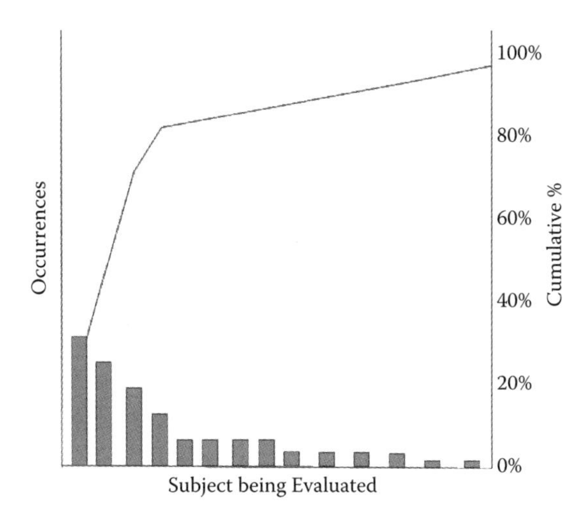

FIGURE 2.6
Pareto chart.

only account for 25% of sales. The Pareto chart could also have been made for defect types or products experiencing failures.

A Pareto chart is an effective tool for determining which quality problem should be addressed first. Generally, either the most common failure or the most costly failure should be addressed first. It may be advantageous to use more than one Pareto chart: one for failure types and one for failure costs. It may be better to address a lower number of failures in an expensive system before a higher number of failures in a simple, low-cost part. Regardless of what the Pareto analysis indicates, safety issues should be addressed first. Sound engineering and economic judgment should be applied when using the Pareto chart.

Another method for visualizing factors is the scatter plot. Scatter plots use paired data comparisons; a treatment variable is plotted on an *x* axis, and a response variable is plotted on the *y* axis. Scatter plots are used to look for a potential correlation between the variables; that is, a change in one variable results in a change in the other variable. For example, a relationship may be suspected between oil pressure and a response variable, so an investigator performs 15 trials and plots the results in a scatter plot. The investigator observes an increase in the oil pressure that corresponds to an increase in response variable, which indicates a correlation may be present. A correlation in a scatter plot is not conclusive evidence of a correlation; the two variables may be moving together in response to a third, unknown, variable (Palady and Snabb, 2000).

An investigator using scatter plots should look for patterns in the points that are plotted in the scatter plot. Points moving upward and to the right may indicate a positive correlation. The indication for a correlation is stronger if an imaginary line is drawn though the center of the cluster of points and the points are close to the line. A high degree of dispersion indicates less chance of a correlation. A negative correlation is like a positive correlation, but going from the upper left side of the scatter plot and moving down toward the lower right side. The data points scattered randomly about the graph are an indication that no correlation is present (Ishikawa, 1991).

A scatter plot can be used for exploring preliminary data or testing a hypothesis that there is a relationship between the variables. A scatter diagram is shown in Figure 2.7. It should be noted that a strong correlation does not necessarily mean that one factor is causing the other. Box and colleagues tell of a strong correlation between the population of the city of Oldenburg and the local stork population over a period of seven years (Box, Hunter, and Hunter, 2005). It may be erroneous to conclude there is a relationship between people and storks.

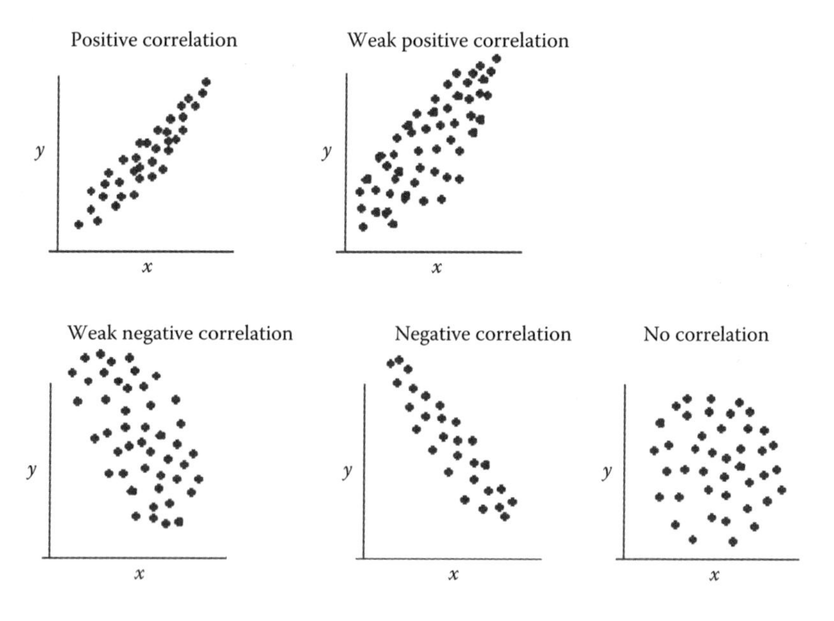

FIGURE 2.7
Scatter plot.

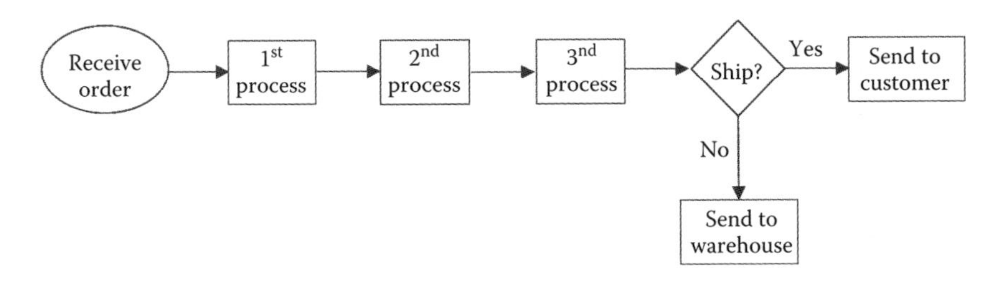

FIGURE 2.8
Flowchart.

A flowchart is used to understand a process (Figure 2.8), unlike the scatter plot and Pareto chart, which are used for data analysis. Also unlike the other classic seven quality tools, the flowchart was not a part of Ishikawa's *Guide to Quality Control*. Regardless of its origins, a flowchart is a useful tool for root cause analysis. A root cause investigator can use a flowchart to map a process and thereby gain a better understanding of the factors relating to the process. For instance, a root cause investigator searching for the root cause of sporadic defective parts from a production machine may create a flowchart of the manufacturing process from the input of the raw material to the completion of the finished parts. The factors listed in the flowchart could then be used for an Ishikawa diagram and then analyzed in detail. A flowchart is particularly useful for process-related failures, such as when a quality manager must determine the root cause for orders being improperly entered into an enterprise resource planning (ERP) system.

A flowchart uses symbols to represent the flow of a process, including activities and decision points in a process. Typically, an oval is used for the start or end of a process, and a diamond is used to symbolize a yes or no decision point. Boxes or rectangles are used to represent the individual activities in a process. Lines with arrows depict the flow of a process. Occasionally, a flowchart may be too long for the paper it is drawn on, so a letter within a circle is used to indicate a break in the flowchart (Brassard and Ritter, 2010). A flowchart can be used for understanding the steps in a manufacturing process.

There is no one correct order for using the classic seven quality tools, and a simple root cause analysis may not require the use of any of the classic seven quality tools. On the other hand, a complex problem may require the use of multiple quality tools to find the root cause. A quality engineer could use a Pareto chart to select the higher-priority problem to investigate and then use a flowchart to understand the process. The factors identified in the flowchart could serve as a basis for an Ishikawa diagram. Once the key factors are identified, a check sheet could be used for data collection directly at the production process, and the resulting data could be analyzed using a histogram.

3

The Seven Management Tools

The seven management and planning tools were the result of operations research in Japan; in the 1970s, the tools were collected and published in one book that was translated into English in the 1980s (Brassard, 1996). The tools can be used to encourage innovation, facilitate communication, and help in planning (Duffy et al., 2012). These are not tools that should be used by one person alone; the great advantage in using these tools comes from using a team approach.

Although these tools are intended for management and planning, they can still be applied during a root cause analysis or when contemplating corrective actions after the root cause has been identified. The tools can be used to graphically illustrate a concept, which makes the concept easier for a team to discuss because all team members can see the ideas being presented. The seven management tools can also be used for the evaluation of potential improvement actions. These tools are no substitute for empirical methods; however, they can be useful in supporting an empirical investigation.

One of the seven management tools is the matrix diagram. There are many types of matrices used for studying the linkage between causes and effects, such as the roof-shaped, L-shaped, Y-shaped, T-shaped, and X-shaped (Wilson, Dell, and Anderson, 1993). The matrix shown in Figure 3.1 is L shaped; it is a simple matrix consisting of one variable listed in the first vertical column on the left and a second variable listed in the top horizontal row. Generally, more than just two variables are used.

The types of variables used should be based on the actual problem under investigation. A matrix may be useful for listing the results of investigations using other tools or the results of hypothesis testing when investigating a complex problem. This ensures that the data collected are not lost or forgotten. Another use for a matrix is to collate data, for example,

	Factor 2				
Factor 1					

FIGURE 3.1
L-shaped matrix.

by listing the characteristic measured in the vertical column and the item measured in the horizontal row. The measurement results would be recorded in the corresponding cells where the measurement characteristic and the measured item meet.

An activity network diagram is much like a PERT (program evaluation review technique) chart, which is used to identify the time required to complete a project (Benbow and Kubiak, 2009) or process (Figure 3.2). This tool can establish the critical path and the tasks that must be performed in a specific order and determine how long a process or project will take. Starting actions on the critical path later than necessary could delay a project; early identification of the critical path can ensure that the tasks on the critical path are known and can be started before tasks that are less critical to the time required to complete the project.

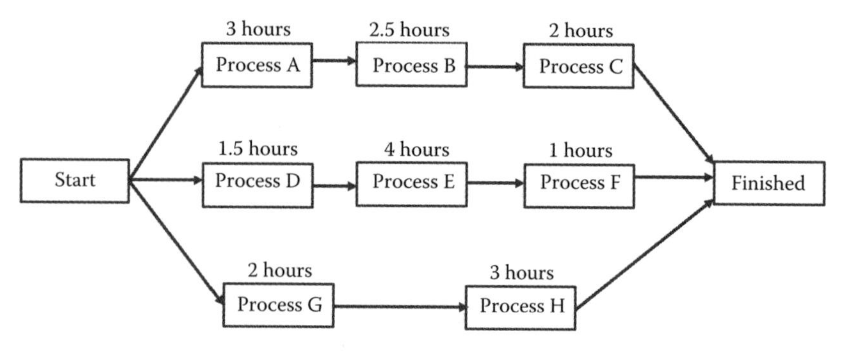

FIGURE 3.2
Activity network diagram.

	Weighted Value	Option 1	Option 2	Option 3
Factor 1				
Factor 2				
Factor 3				
Factor 4				
Factor 5				
Factor 6				
Factor 7				
Factor 8				
	Total:			
	Percentage:			
Degree of fulfillment: 5 = not at all, 10 = medium, 15 = 100% fulfillment.				

FIGURE 3.3
Prioritization matrix.

A prioritization matrix is used to quantify and determine priorities for items (Breyfogle, 2003), such as possible solutions to a problem (Figure 3.3). The options are compared in a table, and factors are given weighted values; this is especially useful if the potential solutions fulfill requirements to different degrees. Using a prioritization matrix helps avoid the selection of a solution that meets all requirements but not well.

Interrelationship diagrams (Figure 3.4) use arrows to show relationships, such as the inputs, outputs, and interconnections among processes (Dias and Saraiva, 2004). This tool can be used to show the relationship between causes and effects and may be particularly useful when evaluating service-related failures involving multiple difficult-to-quantify factors. Arrows are used to show the direction of influences, and one item may influence many other items or be influenced by many other items.

Tree diagrams (Figure 3.5) use a structure with broader categories on top and break them down into details at the lower levels (Liu, 2013). For root cause analysis, the failure under investigation should be listed on top of the tree diagram. The potential causes of the failure are then listed beneath the first failure. A tree diagram can be used to visualize the many

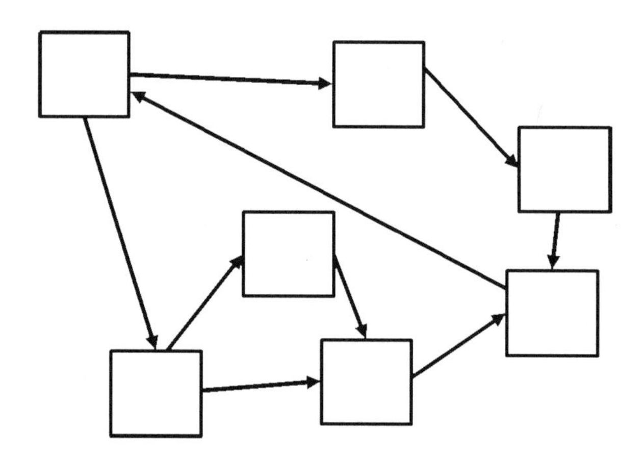

FIGURE 3.4
Interrelationship diagram.

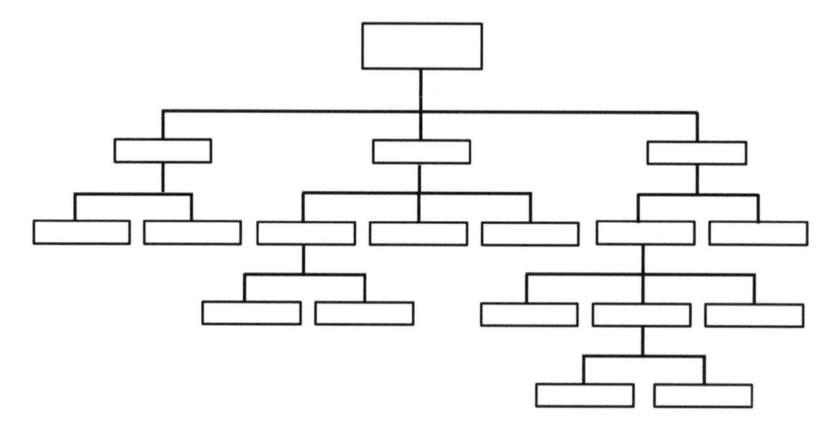

FIGURE 3.5
Tree diagram.

potential failures that led to a final failure; however, the failures identified should still be empirically investigated.

A process decision tree (Figure 3.6) is used when there are many possible solutions to a problem and the best possible solution needs to be selected (Levesque and Walker, 2007). The subject under consideration is listed on the left side of a tree structure, and potential solutions are listed to the right. These solutions may entail potential problems; if so, the problems are listed further to the right. This tool can also be useful when considering multiple improvements, such as when trying to improve a process.

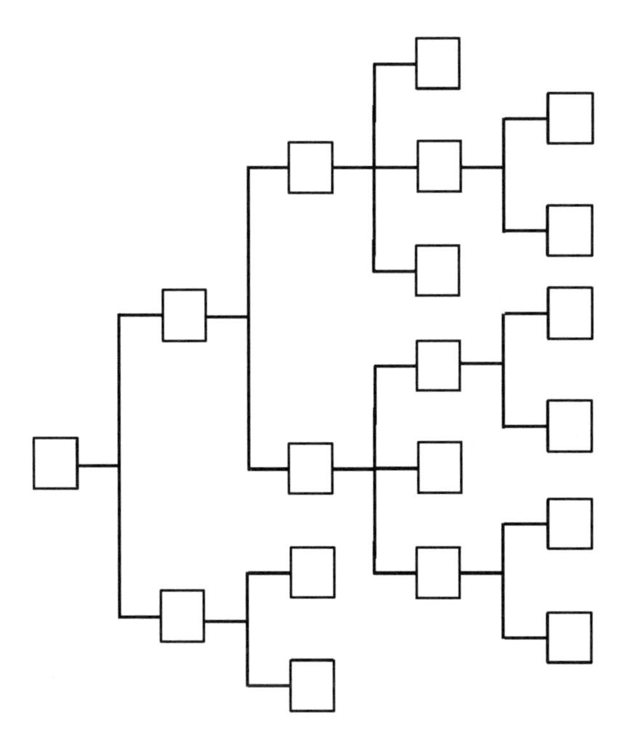

FIGURE 3.6
Process decision tree.

Affinity diagrams (Figure 3.7) group many items in categories, which makes them easier to comprehend (Liu, 2013). Tague (2005) recommends using affinity diagrams when there are many unordered facts, the issue is complex, and a group must reach an agreement, such as when brainstorming. Ideas, such as potential causes or a problem, are written on cards, and then the cards are organized into categories.

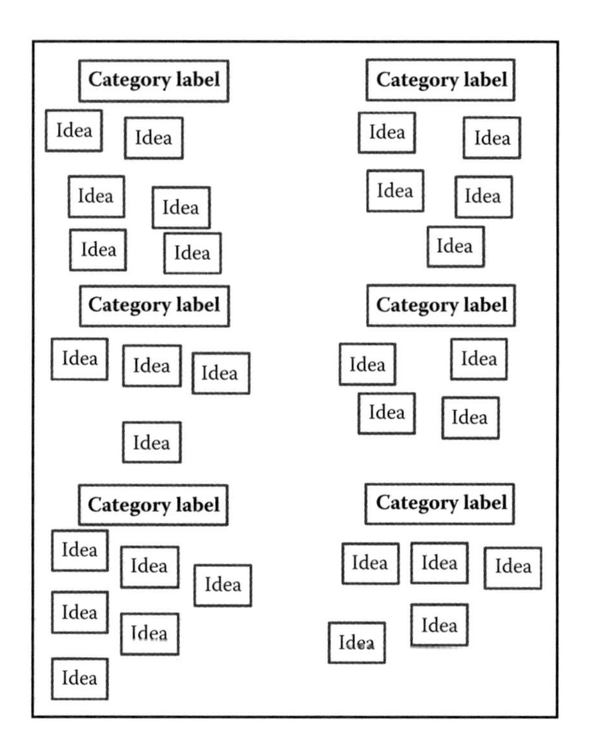

FIGURE 3.7
Affinity diagram.

4

Other Tools for Root Cause Analysis

The classic seven quality tools and the seven management tools are not the only quality tools available to a root cause investigator. Other quality tools may be of great use in assisting a root cause investigator. One simple concept that should be used during any root cause analysis is asking "why" five times. Without asking why more than once, there is a danger that the root cause that has been identified is a contributor to the failure under investigation, but not the true underlying root cause. This could lead to misdirected corrective actions and a recurrence of the failure.

Each failure may have more than one root cause that contributed to the failure. The root cause closest to the occurrence can be labeled the proximate cause (Dekker, 2011). The root cause that resulted in the proximate cause or the chain of events that led to a proximate cause is the ultimate root cause. Using the 5 Why method can lead from the obvious proximate cause to the ultimate cause.

Taiichi Ohno considers repeatedly asking why to be the scientific approach on which the Toyota production system is based. Repeatedly asking why prevents focusing on obvious symptoms while ignoring the true root cause. Ohno presents the following example of the use of 5 Why (1988, p. 17):

1. Why did the machine stop?
   There was an overload, and the fuse blew.
2. Why was there an overload?
   The bearing was not sufficiently lubricated.
3. Why was it not lubricated sufficiently?
   The lubrication pump was not pumping sufficiently.
4. Why was it not pumping sufficiently?
   The shaft of the pump was worn and rattling.
5. Why was the shaft worn out?
   There was no strainer attached, and metal scrap got in.

A root cause investigator who is analyzing the machine stoppage presented in Ohno's example would be mistaken if he or she simply replaced the blown fuse; the failure would occur again. Even lubricating the bearing would only delay the time it takes until the failure reoccurs. The question "Why?" should be asked until it is no longer possible or logical to dig deeper into a root cause.

Imai (1997) thinks that 90% of all quality improvements on the production floor could be implemented quickly if managers used 5 Why. Unfortunately, people often go straight to the obvious answer and do not take the time to use 5 Why. Five Why is a quality tool that should be used during every root cause analysis.

A method that uses comparisons is the is-is not matrix. An is-is not matrix can be used to compile the results of other quality tools. For example, a run chart with stratification may indicate that a problem only occurs on one production line, although the same problem could occur on a second production line. "By comparing what the problem is with what the problem is not, we can see what is distinctive about this problem, which leads to possible causes" (Tague, 2005, p. 330).

Although there are variations in how different authors explain how to create an is-is not matrix, it is generally created by listing statements with questions pertaining to the problem. The questions are then answered in a column for "is" and a column for "is not" as shown in Figure 4.1. The objective is to find a critical difference between where or when the problem occurs and when or where it does not occur but could be expected to occur. This difference may not lead to the root cause, but it does warrant

Problem is	Problem is not	Differences

FIGURE 4.1
Is-is not matrix.

further investigation because it may lead to the root cause. Like other quality tools and methods, this tool should be used with other tools to achieve synergy. As previously mentioned, the data from a run chart could be used in preparing an is-is not matrix. The data collected in a check sheet may also be useful here.

A root cause may not be easy to clearly identify. A root cause investigator may need to see where multiple lines of evidence converge for situations in which the proverbial smoking gun is missing. Using multiple lines of evidence is a technique used by researchers (Beekman and Christensen, 2003) to draw conclusions from multiple pieces of evidence or hypotheses. This convergence of multiple lines of evidence is consilience: a "linking of facts and fact-based theory … to create a common groundwork" (Wilson, 1999, p. 8). The scientist E. O. Wilson explains that this concept was originally presented in William Whewell's 1840 book *The Philosophy of the Inductive Sciences*. Whewell believed the consilience of induction was what happened when the individual inductions resulting from separate facts converge at one conclusion.

A root cause investigator should follow the lines of evidence to the consilience of induction. Naturally, a consilience of induction is not conclusive proof that can be taken as a sign that the root cause has been identified because the rules of hypothesis testing should be followed. Rather, a new hypothesis can be formed as a meta-level hypothesis that should be tested.

A root cause investigator can use other tools, such as run charts, scatter plots, or cross assembling to find evidence that can lead to a consilience of induction. Following lines of evidence is not so much a quality tool for root cause analysis as a concept that can be used to support an investigator during a root cause analysis. In addition to working well with the quality tools, the concept of following lines of evidence can also use statistical data as evidence for hypothesis generation.

A parameter diagram (P-diagram) (Figure 4.2) is often used for creating and deciding between design concepts when using Design for Six Sigma (Soderborg, 2004). Although generally used in product development, a P-diagram can be used during root cause analysis when investigating failures pertaining to a design concept. If there is limited empirical data available, a P-diagram can be used to consider the system as a whole.

A P-diagram considers the inputs into a system, the ideal function of the system, possible error states, control factors, and noise factors. Control factors are what can be controlled, such as quality checks to ensure only parts of the proper dimensions are used. Noise factors are uncontrollable

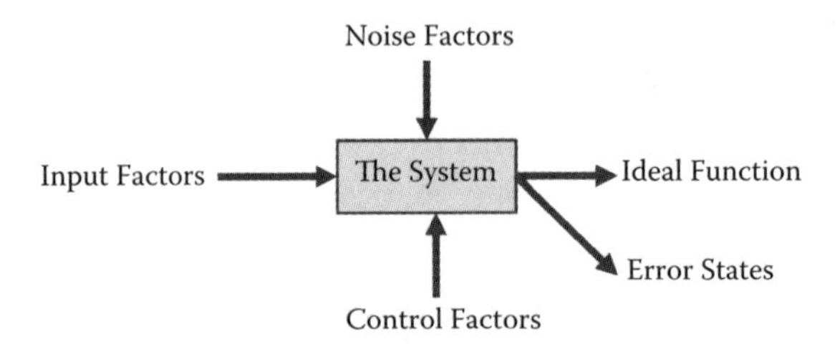

FIGURE 4.2
Parameter diagram.

factors that influence the system; generally, noise factors to consider are the operating environment, customer usage, interactions with other systems, and variation such as between parts.

A boundary diagram (Figure 4.3), also known as a block diagram, is used to depict the relationship between components in a system as well as their interactions. According to the Chrysler, Ford, General Motors Supplier Quality Requirements Task Force, interactions include "flow of information, energy, force, or fluid" (2008, p. 18). The blocks in a boundary diagram represent components, and arrows depict the interactions in

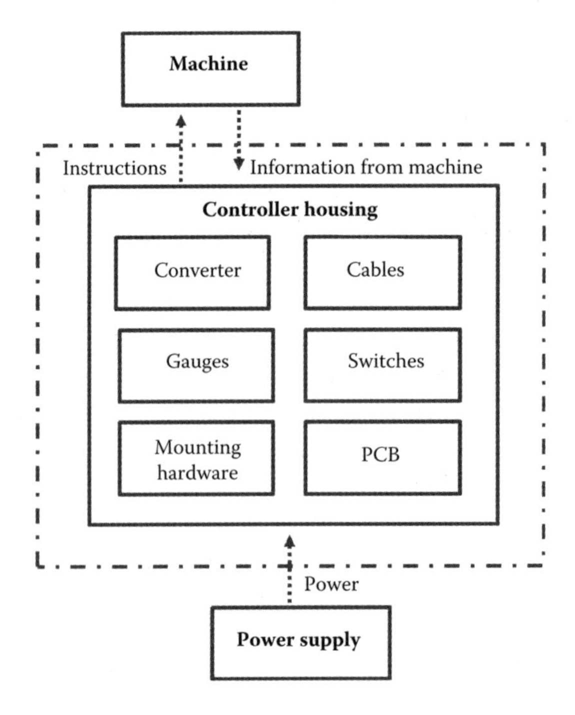

FIGURE 4.3
Boundary diagram.

the system or with other systems. A dotted line surrounding the blocks defines the limits of the system.

Boundary diagrams are often used when creating design failure modes and effects analysis (D-FMEA), for which they can be used to define the limits of the system being considered as well as the system's interactions with other systems outside the limits of the boundary. For example, the brakes on a car must function together with the wheel; however, the wheel would be outside the limits of the brake manufacturer's boundary diagram. The brakes impart a physical force onto the wheel, and this would be represented by an arrow. The hydraulic pressure that engages the brakes would be depicted as an arrow entering from outside the brake's boundaries.

5

Exploratory Data Analysis and Root Cause Analysis

As useful as it may be to use the scientific method coupled with the classic seven and other quality tools, a root cause investigator's toolbox would be incomplete without a method for forming the first tentative hypotheses. Exploratory data analysis (EDA) provides such a methodology. The concept of EDA was created by John Tukey for using statistical methods for hypothesis generation and searching through data for clues. Trip and de Mast (2007) believe "the purpose of EDA is the identification of dependent (Y-) and independent (X-) variables that may prove to be of interest for understanding or solving the problem under study" and EDA can "display the data such that their distribution is revealed" (p. 301).

Tukey calls EDA "detective work" and compares the analysis of data to detective work. During both data analysis and detective work, data and tools are required as well as an understanding of where to look for evidence. Detectives find clues that are later presented to a judge, and an analysis of data seeks evidence that can be confirmed through more rigorous testing at a later time (Tukey, 1977, p. 1). Many tools that could be used during EDA may not be specific EDA tools; for example, the classic quality tools could be used as a part of EDA to graphically view data. The data are then searched for "salient features" (de Maast and Trip, 2007, p. 369), that is, features or characteristics that do not conform to what would be expected or that stand out as different from expectations.

Trip and de Mast (2007) illustrate the use of EDA in a situation for which they used a Pareto diagram to determine which workstation generated the most defects during the assembly of motors; the Pareto diagram led them to one workstation that was different from the others and producing the majority of defective motors. They then also recount the case of eccentricity of pins in cell phone production. A histogram was constructed using final inspection data, and the histogram depicted bimodal data, that is, two bell-shaped distributions. This led the investigator to suspect that there were two distinct populations present in the data. The operators confirmed that parts came from two different molds; however, it was not possible to determine which parts came from an individual mold. New data then confirmed the hypothesis that the variation was a result of the molds.

Tukey's EDA is not any one specific method but rather a collection of tools used for the analysis of data. One of these tools is the stem-and-leaf plot. A stem-and-leaf plot is used to visually display a data set containing numbers with two or more digits. The first column is the stem; the leaves are the rows to the right of the numbers on the stem. The numbers are organized from lowest at the top to highest at the bottom (Montgomery, Runger, and Hubble, 2001). The first digit or digits in a number is placed in the stem, and the remaining digit is placed in the leaf. Using a data set containing three-digit numbers would require placing the first two digits of each data set in the stem and the remaining digit in the leaf. Using the numbers 112 and 115, a stem would contain 11 and the leaf would contain 25. A legend should be used to clearly identify the units in the stem and leaf as shown in Figure 5.1.

Stem-and-leaf plots can show the frequency distribution of the data; they are much like a histogram. The advantage in using a stem-and-leaf plot is that it is a quick-and-easy tool to use by hand, unlike the histogram, which can be hand drawn but is much easier to create using a graphics or spreadsheet program.

A more informative method for displaying data is a box plot. A box plot is constructed with a box using the limits of two quartiles as the ends of

Stem	Leaf
56	269
57	24566
58	0135788
59	4567899
60	124568889
61	13345667789
62	0133557
63	13699

FIGURE 5.1
Stem-and-leaf plot.

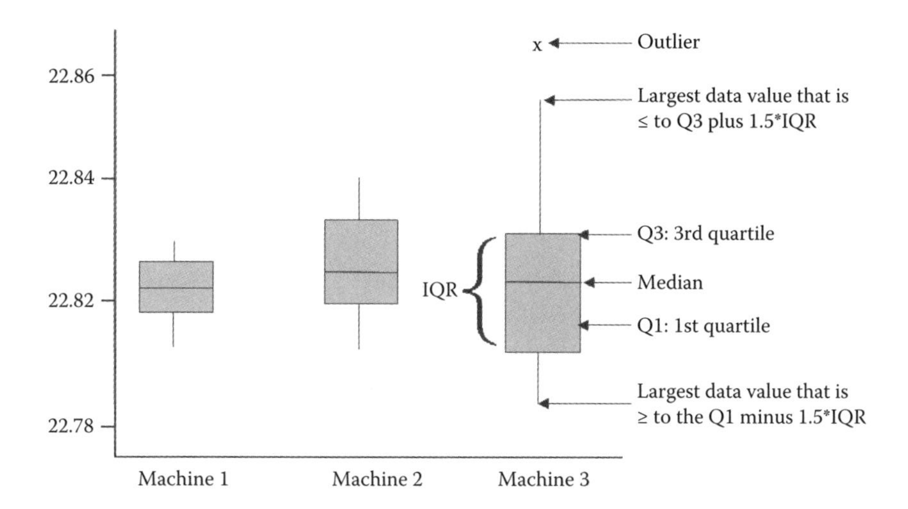

FIGURE 5.2
Box plot structure.

the box and a horizontal line in the box to identify the median. Whiskers originating at the box ends are used to display the spread of the data away from the median (Gryna, 2001). The upper whisker's length is determined by drawing a line from the upper box end to the largest data point that is less than or equal to the third quartile value plus one and a half times the interquartile range. The interquartile range is the distance between the second and third quartiles. Outliers in the data set are identified by an *x* (George et al., 2005), such as the one seen in Figure 5.2.

Box plots are a quick-and-effective method for comparing data sets. This could be handy in root cause analysis for comparing the difference in the spread of data between machines or parts. A box plot displaying multiple data sets can be seen in Figure 5.3, and the same concept can be used for process-related data.

Multi-vari charts are another graphical tool used to study variation. Multi-vari charts are used to show how the variation in an input variable affects an output variable. Data for a multi-vari chart can either be pre-existing or generated though experimentation. Sources of variation can be the variation between parts, variation over a predetermined length of time, or cyclical variation. A sampling plan should be created to graphically display the data sets that will be used (Sheehy et al., 2002). Observations from a multi-vari chart can serve as a catalyst for hypothesis generation, which is the objective of EDA. An example of a multi-vari chart is shown in Figure 5.4.

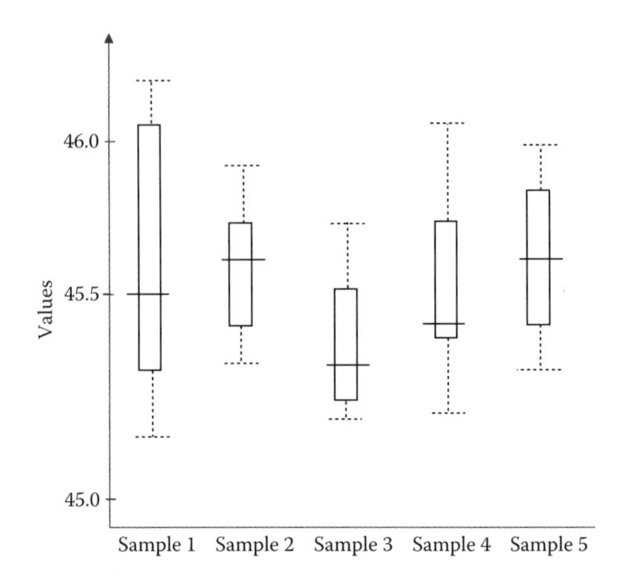

FIGURE 5.3
Box plot.

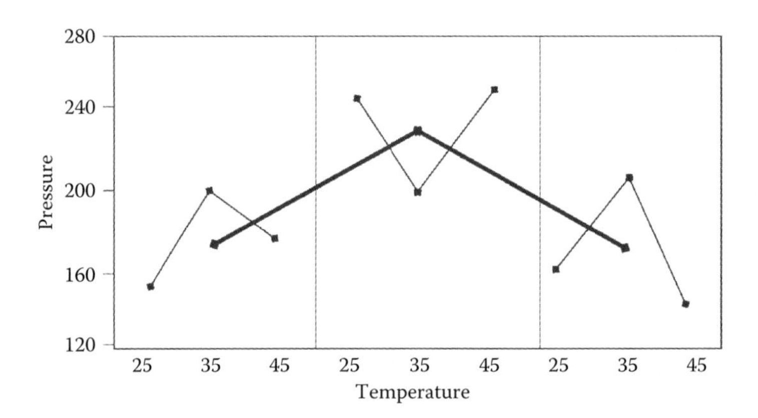

FIGURE 5.4
Multi-vari chart.

Hypothesis generation can be greatly assisted by EDA. Kemper and de Mast (2009) use the 1854 London cholera outbreak to demonstrate the uses of EDA in hypothesis generation. Some parts of the history of the investigation of the cholera outbreak may be apocryphal; that is, legend and truth pertaining to the investigation details may have blended. In spite of this, the story of the investigation can be used to illustrate EDA in action.

A simplified account of events holds that during the 1854 cholera outbreak a man named John Snow suspected a waterborne cause for the outbreak, which was contrary to the then-current theory of the origins of cholera. Snow plotted the homes of cholera victims on a map and determined which of two water companies was supplying the drinking water used by the homes. Most of the affected homes were found to have received their water from a pump on Broad Street, so Snow removed the pump handle and ended the outbreak. Some data conflicted with Snow's hypothesis: A factory and brewery near the pump were relatively unaffected, and 10 victims lived away from the others and the Broad Street pump. Snow investigated and determined that the factory and brewery had their own wells, and many of the people who died outside the affected area either passed by the Broad Street pump and used it or preferred it and sent servants to fetch the water. Further investigation determined a baby in a house near the Broad Street pump had died of cholera just before the start of the outbreak; the baby's diapers were washed and the mother dumped the water into a drain next to the Broad Street pump (de Mast and Kemper, 2009). Snow used the principles of EDA over a hundred years before the concept was named. He visually depicted his data and looked for patterns in the data for the generation of a hypothesis that could be supported by empirical data.

6

Customer Complaint-Related Root Cause Analysis

A complicated root cause analysis requires more than just the application of quality tools and a thorough analysis. Actions must be tracked, and data need to be consolidated. A folder on a server with the investigation's reports and a tracking list for action items and results may be sufficient for a simple investigation in a small company.

Immediate actions must also be considered if the root cause analysis is because of a quality failure. It is not good to ship hundreds of defective components to the customer while investigating a failure. The exact nature of the containment actions should be based on the issue and could vary from random sampling to 100% inspection of parts in production, inventory, transit, and at the customer's location. The PDCA (Plan-Do-Check-Act) cycle can also be used for the start of an issue (Figure 6.1).

Lessons learned should be considered at the end of a root cause analysis. Other improvement opportunities should be sought if a product or process is improved based on actions resulting from the investigations. Lessons learned during the investigation should be saved so that the same problem is prevented from occurring at other locations in the future. For a small company, this could be as simple as recording the lessons learned in a spreadsheet that is periodically reviewed, such as when designing a new product or process.

Recording lessons learned is more complicated when dealing with large, multinational corporations. The proper recording of lessons learned may also be more important for large corporations for which the same part is produced or the same process is used in many locations. Those involved in the root cause analysis are not available to remind others of what they have learned. One option for large companies is a central lessons learned

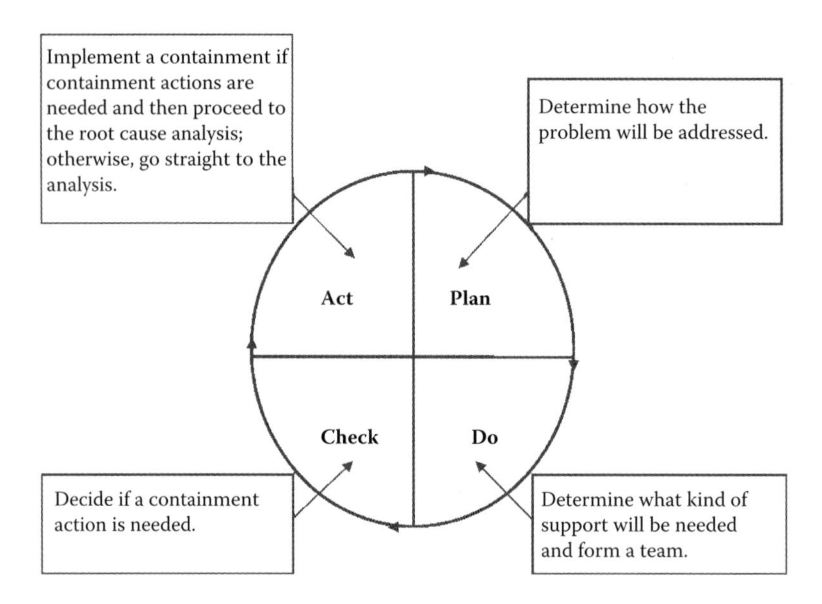

FIGURE 6.1
PDCA for the start of an issue.

database. Another option is to ensure all design standards, company standards, or standard documents, such as failure modes and effects analysis (FMEA), are updated based on the lessons learned. The exact nature of the documentation of lessons learned must be based on the needs and requirements of the company.

For root cause analysis resulting from quality failures, an alternative approach to a simple tracking list is to use an 8D report. An 8D report is not a quality tool but a method for addressing and investigating quality failures. The 8D report will not identify the cause of a failure, but it will provide a structure that requires immediate actions and preventive actions to be carried out. It also provides an easy-to-understand report on the investigation that can be saved as a form of lessons learned after the issue is closed.

The predecessor to the 8D report was *MIL-STD-1520C: Corrective Action and Disposition System for Nonconforming Material*, which was created by the US Department of Defense in the 1940s for dealing with quality problems in products supplied to the Department of Defense (1986). The concept of the 8D report was further developed by Ford Motor Company in the 1970s and quickly spread through the automotive industry and then outside the automotive industry.

Report No.: Customer Claim No. Supplier 8D No.:	Part No.:
	Complaint:
(1) Team	Opened on: Version date:
(2) Problem Description	
(3) Immediate Containment Action	Resp./Date
(4) Root Cause	
(5) Planned Corrective Action	Resp./Date
(6) Implemented Corrective Action	Resp./Date
(7) Actions to Prevent Reoccurrence P-FMEA D-FMEA Control Plan Procedures	Resp./Date
(8) Congratulate the Team	Closed on

FIGURE 6.2
Example of an 8D report.

Typically, 8D reports are used for reporting to customers on supplier failures. A company that receives a customer complaint documents its activities in the 8D report. The same company can require an 8D report if it receives defective parts from a supplier. Although 8D reports are often used when dealing with customer complaints, they can also be used for internal issues, such as when a defective part from one process is detected at a further process or at a final inspection.

The *D* in 8D stands for "disciplines"; these disciplines are the steps in an 8D report (Figure 6.2). The name of the actions at each step may vary from one company's 8D report to another company's 8D report; however, the actions to take at each step should not vary between companies. The actual 8D form could be in the form of a text document, a spreadsheet, a slide-based presentation, or a document generated by a company's computer system. The eight steps are as follows (Rambaud, 2011):

D1: Use a team approach.
D2: Describe the problem.
D3: Implement and verify the temporary fix.
D4: Use root cause analysis.

D5: Develop permanent solutions.
D6: Implement and validate a permanent solution.
D7: Prevent reoccurrence.
D8: Close the problem and recognize contributions.

Prior to step D1, the form's header information should be completed. This varies between companies and should fit the needs of the company using the 8D report. Generally, a number should be assigned to the 8D report, and this number is then listed on the document. For customer complaints, the customer's complaint number should be listed on the 8D report. This makes it easier to tie the company's 8D report to the customer's issue. It also makes it easier for the customer to determine to which issue the 8D report pertains. The supplier's 8D report number should also be listed if the issue is the result of a supplier failure. The date the complaint was opened needs be listed as well as the date of any updates. The company's part number should be listed in the 8D report. Listing the customer or supplier's part number will make working the issue easier. Customers often insist on using their part number; a company that lists both on the 8D report can quickly identify the part to which the customer is referring. Listing the supplier's part number is also helpful. The complaint should be given a short and descriptive name, such as such as "crack at elbow joint." Additional information fields should be added based on the needs of the company using the 8D report.

The first step is forming a team. The team must be interdisciplinary and have representatives from all departments that are affected by the issue. There should be a designated team leader who ensures that all assigned activities are carried out and the 8D report is regularly updated. There should also be a champion; this is a person in a management position with sufficient authority to assist the team if the team encounters difficulties or needs additional resources.

The second step is describing the problem. This should be done in terms of what is relevant to the customer, such as "customer reports a part has been found in their production with a crack at the elbow joint." This is not the right place to list the root cause of the failure; that should be identified later and only after an investigation.

The next step is the immediate containment action. This step may not always be necessary, such as when only one unit has been produced and the customer will use it as it is. However, this step should always be considered, and a brief explanation should be given if it is determined not relevant. If containment is required, the person responsible and the implementation date should be given. It may also be necessary to explain how checked parts will be identified so that the customer will know when good parts arrive; otherwise, the customer may be checking all incoming parts even after the supplier has determined they are good. The number of parts checked and the inspection results can be recorded here. The first 8D report should be sent to the customer once this step is completed.

A root cause analysis, requiring the use of quality tools should be performed and described in detail in step 4. A thorough description of the investigation actions helps to reassure the customer that the supplier has the situation under control and is progressing toward the identification of the cause of failure. This step may require an additional sheet of paper for details or the attachment of documents, such as laboratory or measurement reports. An interpretation and summary of the reports should be available in the 8D report. Many companies simply list "worker error" when the problem was caused by an employee's mistake; this is a poor choice in root cause analysis. It is better to describe how and why the person made the mistake. The root cause of the failure to detect the problem should also be described.

The planned corrective actions are detailed in step 5. Here, the planned actions as well as the method or methods used to verify that they will actually work should be described. A common proposed action for human error is "employee training." Perhaps training is necessary; however, this does little to ensure the failure cannot happen again. It is better to implement changes that will prevent the possibility of a failure. For example, if somebody forgets to install a component, it may be necessary to design an automatic check device to ensure that the assembly cannot leave the workstation if the component is missing. Corrective actions should also consider ways in which a reoccurrence of the failure can be detected prior to it reaching the next process or the customer.

Step 6 is the implementation of corrective actions that have been evaluated and found effective. Although these actions have been evaluated in the previous step, they should still be monitored to ensure that they are effective after implementation.

An additional step is taken to ensure the failure cannot occur again. This is often done by updating documents such as the FMEA and control plan as well as standards and procedures. This step is not only to ensure the failure will not reoccur on the same part that previously failed but also to ensure that other parts or processes cannot experience the same failure.

The final step is congratulating the team to ensure the team understands that its contributions were appreciated. The 8D team is disbanded at this point, and a new team is formed if a new failure occurs. The new team may consist of the same team members or others if different departments are affected by the failure. The 8D report is closed at this point and submitted to the customer or department that initiated the complaint. The closed-out 8D report should then be filed in a location where it can be accessed if it is needed as a lesson learned.

7

Example of a Root Cause Analysis

Root cause analysis should be a mixture of the scientific method and the necessary tools for collecting and analyzing data as well as hypothesis generation and evaluation. Suppose a hypothetical quality engineer decides to look for improvement opportunities, so he looks into the customer complaint database for the past year. The quality department has limited time to dedicate to improvement activities, so the quality engineer uses the customer complaint data to create a Pareto chart to prioritize issues. The customer complaints data indicate that the top three problems are rust, length deviations, and diameter deviations. Rust complaints account for 55% of all complaints, so the first quality improvement will concentrate on rust.

The quality engineer uses PDCA (Plan-Do-Check-Act). The plan phase consists of looking into the customer complaint data and using stratification to separate the complaint characteristics; a run chart is created. The run chart indicates that there are more complaints for smaller-diameter parts with rust than larger-diameter parts. The quality engineer uses deduction to hypothesize that more small-diameter parts were sold, and that is the reason for the higher number of complaints. The do phase consists of a quick check of completed work orders for the past year; these indicate the quantities are approximately the same. The check of the work orders results in rejecting the hypothesis and forming a new hypothesis.

Induction is then used to hypothesize that smaller-diameter tubes rust before larger-diameter tubes, and an experiment is planned to investigate the hypothesis. The production department manager is consulted and explains that there are two types of steel that are used. Performing an experiment without controlling the variables could result in confounding, so the quality engineer randomly selects three different small-diameter tubes and three different large-diameter tubes. All of the tubes that are selected were produced from the same type of metal and the same steel coil.

The quality engineer interviews the machine operators and discovers the level of rust inhibitor in the rust inhibitor solution can vary from day to day. He decides to use a large bottle of rust inhibitor for the experiment, and the bottle is to be tested before each experiment to ensure that the rust inhibitor concentration is always 67% with a tolerance of plus or minus 1%.

The six samples are cleaned to remove potential contaminants and then sprayed with rust inhibitor. They are placed in an environmental chamber and checked every day at noon to see if rust is present. All six samples are found to be rusty on the third day. The quality engineer then repeats the procedure using the other type of steel and obtains the same results after three days. The empirical data do not support the hypothesis that smaller-diameter parts rust before larger-diameter parts.

The person responsible for the investigation gathers representatives from engineering, logistics, and production to brainstorm and create an Ishikawa diagram listing factors that could influence rust. These factors are used for an is-is not analysis. The quality engineer attempts to identify salient features and differences between large- and small-diameter tubes. The production processes for both parts were almost the same, with diameter as the only noticeable difference. The is-is not analysis indicated that tube packaging varies between large- and small-diameter tubes. The bundle sizes were the same, but smaller tubes were far more numerous in a bundle.

The warehouse department was instructed to inform the quality engineer if any rusted parts were found in stock. A few days later, he was called to the warehouse to view a bundle of rusted tubes. The quality engineer observed heavy rust on the outside tubes but no rust on the inside tubes. The inside tubes had been protected from the atmosphere by the outside tubes. The quality engineer then hypothesized small- and larger-diameter tubes rusted at the same rate, but more numerous small tubes were exposed to atmosphere when packed in a bundle; therefore, more numerous small tubes were reported as rusted. Placing small bundles of both small- and large-diameter tubes in the environmental chamber supported the hypothesis; outside tubes rusted before inside tubes, and there were more outside tubes in a bundle of small tubes.

The root cause analysis team then hypothesized storage location was a factor in rust formation. This was just a tentative hypothesis that the quality engineer expected to reject; to investigate it, the quality engineer hung samples consisting of large and small tubes of each steel type on the warehouse pillars. He then created a sketch of the warehouse layout and placed an *X* over the location of samples found with rust. Within a week, it was clear that the rust was occurring in two locations, both near openings that led to loading docks. The new hypothesis was "rust occurs primarily near the loading docks," and verification testing was needed; a new trial was performed by hanging samples on all pillars with the expectation that those nearest the loading docks would be the first to rust. The hypothesis received additional support when it was determined that bundles near the doors rusted more often than bundles in the middle of the warehouse. The samples near the loading docks were the first to rust, so the next step was narrowing down the root cause so an improvement action could be implemented. A visual check of the loading dock doors found they were always open; the quality engineer hypothesized that this left the parts nearest the doors more exposed to moisture from the outside environment.

The data on hand were analyzed with the 5 Why method: Small tubes rust more often than larger tubes. Why? There are more tubes exposed to atmosphere in a bundle containing smaller tubes. Why are the tubes rusting? The tubes are rusting because they are near an open door. Why does being near an open door cause tubes to rust? Tubes near an open door rust because the doors are always open, exposing the tubes to moisture from outside. Why are the doors always open? The doors are always open so forklifts can easily drive through. Why are the doors always open so the forklift drivers can drive through? They are open so the forklift drivers will not lose time by stopping to open the doors. Therefore, the root cause is loading dock doors being open all the time so that forklifts can save time when passing through the doors and the open doors are exposing the tubes to moisture from outside.

The problem was discussed with the warehouse manager, and in this situation, the confirmation test was the same as the corrective action: hanging heavy, clear plastic strip curtains so that the warehouse was not directly exposed to the outside but forklifts could easily drive in and out. The samples near the doors no longer rusted, and the complaint rate for rust dropped. The quality engineer then thanked the team for its support and moved on to the next issue.

Not every quality tool needs to be used during a root cause investigation, as illustrated by this hypothetical root cause investigation. The goal should be to select and use the appropriate tool for the situation. The objective of the first hypothesis should not be to immediately zero in on the root cause but rather to eliminate a possible root cause and to gather new data that lead one closer to the true root cause. It is better to move through many quick iterations of induction and deduction using PDCA together with the scientific method. Once a root cause has been identified, a root cause investigator should confirm that the root cause is the true root cause. Corrective actions can then be implemented to prevent a recurrence of the failure.

Section II

Root Cause Analysis Quick Reference

8

Introduction to Root Cause Analysis

INTRODUCTION

Although the formula $y = f(x)$ may be primarily well known to Six Sigma practitioners, it is also applicable to root cause analysis (RCA). The y stands for an output, and the $f(x)$ represents an unknown factor that influences the output y. The formula $y = f(x) + \xi$ (Box, Hunter, and Hunter, 2005) is perhaps more appropriate to RCA with the ξ (named ix) representing error. The objective of RCA is to solve for x while being careful to avoid or account for error.

Root cause analysis is the process of searching for a root cause, the x of the previous formula. The root cause could be the cause of a failure or a condition, such as the reason for the actual output of a process. Each root cause investigation is different, so it is not advisable to seek one method for every situation; rather, a toolbox containing many different tools to use as appropriate is a better approach to RCA.

A failure or unwanted occurrence may have both a proximate cause and an ultimate cause. A proximate cause is the immediate cause and generally easier to identify than an ultimate cause. A proximate cause could be a temporary worker forgetting a process step; the ultimate cause is a work instruction that did not contain the step. An RCA investigator should seek the ultimate cause.

There are commonalities between different root cause investigations; a team approach should be used as well as the scientific method and quality tools. A cross-functional team should be used (Gryna, 2001); however, the size and type of team as well as the tools should be appropriate to the complexity of the issue under investigation. A larger, more formal team may be appropriate for complex investigation. Seeking the advice of subject matter experts may be sufficient in a simpler investigation. The choice in team members or individual subject matter experts should be based on the subject under investigation. The failure of a part in production may only need the assistance of the machine operator and the design engineer; a failure in a logistics system may require purchasing, sales, logistics, machine operators, a forklift operator, and production planning.

Key Points

- RCA can be performed to solve a problem or to identify improvement opportunities.
- Teams should be used, ideally consisting of representatives from different departments.
- An event or failure could have both a proximate and an ultimate root cause.
- Flexibility is needed in the selection of tools.

9

The Science of Root Cause Analysis

HYPOTHESIS AS A BASIS FOR NEW KNOWLEDGE

Forming and evaluating hypotheses is a basic part of the scientific method and should be an essential part of root cause analysis (RCA). A root cause investigator should form a hypothesis to account for the evidence on hand or collect new information to form a preliminary hypothesis. Then, the hypothesis should be empirically checked; the resulting observations serve as a basis for either a new hypothesis or confirmation testing of the hypothesis.

A hypothesis is like a well-reasoned guess and is generally tentative; an attempt must be made to empirically support the hypothesis or to refute it. It should have the following characteristics: conservatism, modesty, simplicity, generality, and refutability (Quine and Ullian, 1978). A hypothesis can never be proven, only rejected or merely corroborated. A hypothesis with a high degree of corroboration is well supported; a hypothesis that has survived many tests is more corroborated and therefore stronger than a less-well-supported hypothesis.

A hypothesis can be formed by inductive reasoning or deductive reasoning. Induction uses specifics such as observations to reach a general conclusion to account for the observations. Deduction uses general prior knowledge to reach a specific conclusion (Box, Hunter, and Hunter, 2005).

An ideal hypothesis will make assumptions and should not diverge too widely from what is already known; a hypothesis that makes too many casual connections is weaker and at more risk of rejection than a simpler hypothesis.

Key Points

- A hypothesis should not make too many assumptions and should be simple and general. It must make a testable prediction to be of any use.
- Occam's razor is a rule of thumb that states the hypothesis with the fewest assumptions should be selected when confronted with two competing hypothesis.
- It is not possible to prove a hypothesis, only to support a hypothesis or reject a hypothesis. A well-corroborated hypothesis is one that has survived many attempts at refuting it.
- Deduction goes from the general to the specific; deductive reasoning uses what is known to form a hypothesis. It uses previous hypotheses and models to form a new hypothesis.
- Induction goes from the specific to the general; inductive reasoning uses observation to form a hypothesis. It uses facts and the observation of phenomena to form a hypothesis.

EXAMPLE 9.1

Deduction: A quality manager observes that every time the pressure in a machine drops, defective parts are produced. On being informed that the pressure has dropped, the quality manager uses deductive reasoning to form the hypothesis: "Defective parts were produced during the time when the pressure was low." The quality manager then begins checking the parts to support or reject the hypothesis based on empirical data.

Induction: A quality engineer is aware that defective parts often result from low pressure in a machine. The quality engineer is informed that defective parts have been found in final inspection and uses inductive reasoning to hypothesize: "Defective parts are the result of low pressure in the machine." The quality engineer then checks the pressure in the machine to support or reject the hypothesis.

PROCEDURE

Step 1: Observe or collect the relevant data.

Step 2: Analyze the data and look for relationships or patterns in the data.

Step 3: Use the data to form a general hypothesis that goes beyond the data to make a prediction.

Step 4: Test the hypothesis.

Step 5: Use the test results as a basis for a new hypothesis if they fail to support the hypothesis. Perform confirmation testing if the results support the hypothesis.

SCIENTIFIC METHOD AND ROOT CAUSE ANALYSIS

The scientific method is a time-tested method of using hypotheses to gain and evaluate information; in RCA, this information leads to the root cause of an event or occurrence. According to de Groot (1969), the scientific method uses an empirical cycle consisting of observation, induction, deduction, testing, and evaluation.

This corresponds closely to Box's iterative inductive-deductive process (Figure 9.1), by which ideas such as hypotheses, models, and conjecture use deduction to reach a tentative conclusion that is then checked against data such as facts and phenomena. The results of the data are then used as a basis for induction, which leads to a new hypothesis (Box, Hunter, and Hunter, 2005).

The scientific method can be used in combination with Plan-Do-Check-Act (PDCA) (Figure 9.2), also known as the Deming or Shewhart cycle. During the plan phase, observations are made or data are collected. The actual actions to take are dependent on each individual situation. The observations or data are used to form a tentative hypothesis. The hypothesis is tested during the do phase; it could be tested by an experiment or making observations. The observations or experimental results are checked and compared against the predicted results during the check phase. The final phase of PDCA is act; here, actions are taken based on the results of the previous phase. Confirmation testing is needed if the results supported the hypothesis; a new cycle of PDCA begins if the result failed to support the hypothesis. Ideally, each iterative of the PDCA should bring a root cause investigation closer to the root cause either by eliminating possibilities or moving closer to the underlying true state of nature.

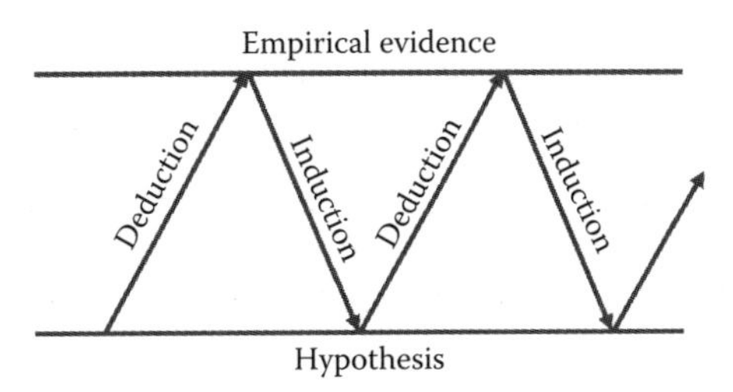

FIGURE 9.1
Box's iterative inductive-deductive process. (Adapted from George E. P. Box. *Journal of the American Statistical Association* 71 no. 356. (1976): 791.)

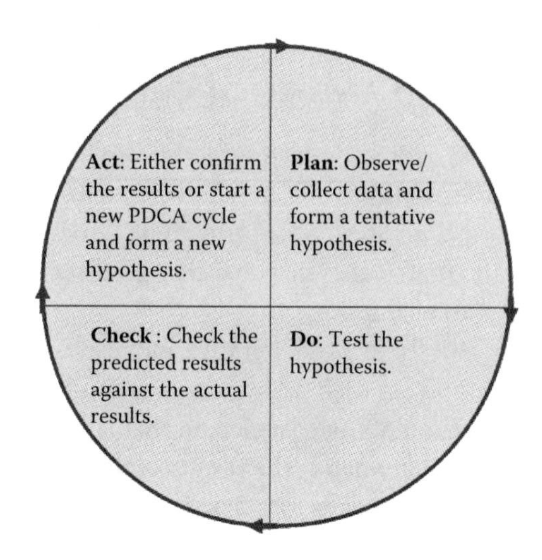

FIGURE 9.2
Scientific method with PDCA.

The objective in using the scientific method to form and evaluate hypotheses in the search for a root cause should not be to immediately find the root cause but rather to provide a systematic method for progressively moving closer to a hidden root cause.

Key Points

- The scientific method is empirical/evidence based.
- A hypothesis is a key part of using the scientific method.
- Quick iterations of the iterative inductive-deductive process can lead to the root cause.
- The objective should not be to immediately identify a root cause but rather to move quickly through the cycles of PDCA to reject unsupported hypotheses and arrive at the root cause.
- Avoid holding on to a pet hypothesis once it is unequivocally disproved. Evaluating an incorrect hypothesis is not bad; defending one is.

EXAMPLE 9.2
This example illustrates the scientific method with PDCA (Figure 9.3).

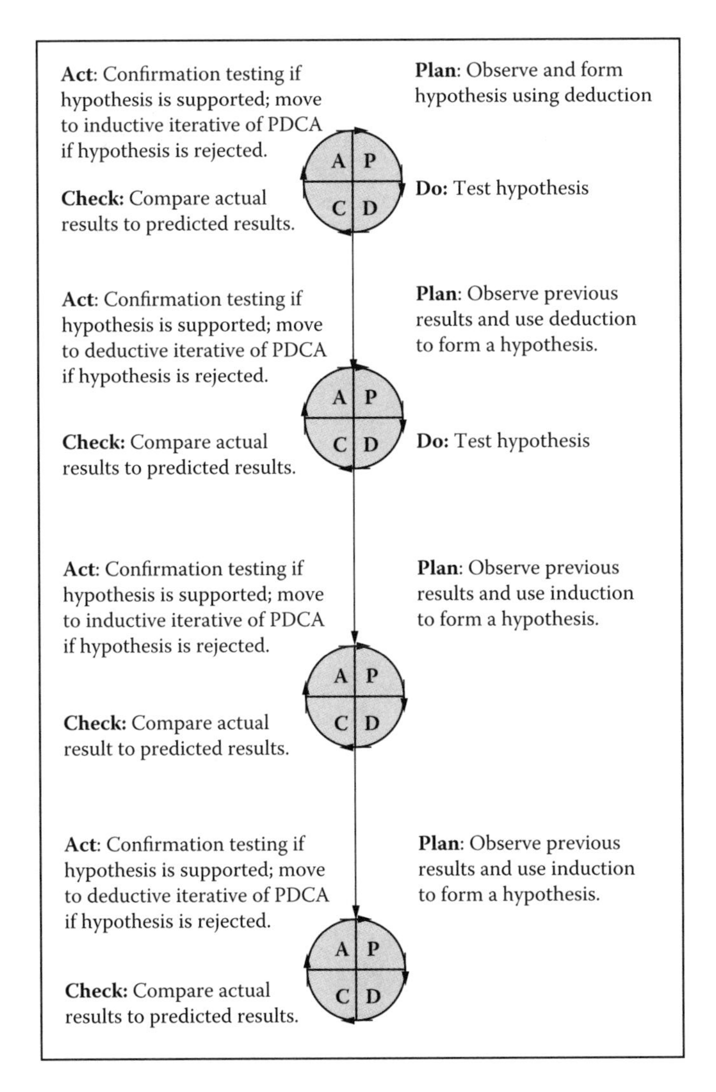

FIGURE 9.3
Multiple iterations of the scientific method with PDCA.

PROCEDURE

Step 1: Plan: Use deductive reasoning to form a tentative hypothesis based on observation.

Step 2: Do: Assume the hypothesis is true for the sake of testing and evaluate it empirically, through either experimentation or observation.

Step 3: Check: Compare the actual results against predicted results.

Step 4: Act: Perform confirmation testing if the hypothesis was supported or move to the next cycle of PDCA and use induction to form a new hypothesis if the tested hypothesis was not supported.
Step 5: Plan: Use the knowledge gained from the previous PDCA cycle to form a new hypothesis using induction.
Step 6: Do: Assume the new hypothesis is true for the sake of testing and evaluate it empirically, through either experimentation or observation.
Step 7: Check: Compare the actual results against predicted results.
Step 8: Act: Perform confirmation testing if the new hypothesis was supported or move to the next cycle of PDCA and use deduction to form a hypothesis if the tested hypothesis was not supported.
Step 9: Repeat the iterations of the PDCA cycle until a hypothesis is supported, at which time confirmation testing should be performed.

EXPERIMENTATION FOR ROOT CAUSE ANALYSIS

Some hypotheses can be supported or rejected by a simple observation to see if they fit facts. Others may require experimentation to evaluate them. A treatment is the set of conditions in an experiment. During an experiment, the treatment variable is a factor that is manipulated or adjusted by the experimenter to determine its effect on the response variable, which is the outcome of the experiment. Confounding variables or noise are variables that can affect the results of an experiment but are uncontrolled. Variables should be controlled whenever possible or otherwise dealt with when not controllable. One method of dealing with confounding variables is blocking. An uncontrolled variable is distributed across all treatments by ensuring all experimental groups or blocks contain the confounding variable, thereby decreasing variability and increasing the precision of the experimental results. Box advises to "block what you can and randomize what you cannot" (Box, Hunter, and Hunter, 2005, p. 93).

Randomization can cancel or minimize the effects of noise and thereby increase the accuracy of experimental results. To randomize, samples should be randomly selected, such as taking them from multiple levels in a box and not from the top; it is possible that something changed during the production process and the parts on the top of the box are not representative of the entire population. Replication is also used to increase the accuracy as well as the precision of an experiment. Repeating an experiment can provide better data by including variation in the results. A treatment that is not replicated may be influenced by variation and does not provide as accurate a picture of the true results as multiple experimental runs.

The precision of the results is their closeness to each other across multiple measurements or experiments. Accuracy is the closeness of experimental or measurement results to the true value. The ideal is both precision and accuracy. An experiment could produce results with high precision and low accuracy; the results would be consistent, but wrong.

An experimenter should also use operational definitions, which are quantitative descriptions of terms that are used so that the meanings of the terms are unambiguous. Operational definitions must be clear enough for anybody using them to understand and should use measurements or tests to define the terms (Deming, 1989).

It may be advantageous to establish a baseline, the output of a process prior to changes being made. An experiment may indicate that changes to the treatment variable resulted in a change to the response variable; however, it is also possible that the response variable would have changed because of an unknown noise factor regardless of the setting of the treatment variable. The experimental results should be compared to the baseline.

Key Points

- A treatment, also known as an experimental run, is the set of conditions during an experiment.
- A factor is a condition that affects an output, for example, temperature, material type, mixture, settings on a machine, or pressure.
- The treatment variable, also known as an independent variable, is a factor that is manipulated by the experimenter to determine its effect or lack of effect on the response variable.
- A response variable, also known as the dependent variable, is the result of the manipulation of the treatment variable.
- The confounding variable, also known as the confounding factor, is a source of noise.
- Noise in an experiment is an uncontrolled and potentially unknown factor that influences the experimental results.
- Precision is the closeness of measurements to each other.
- Accuracy is the closeness of a measurement to the true value.
- Blocking reduces variability and increases precision by spreading confounding variables across the experimental results.

- Randomization increases the accuracy of experimental results by canceling out the influence of noise.
- Replication is the repetition of an experiment to increase the accuracy and precision of the results.
- Operational definitions are clear quantitative descriptions of terms using tests or measurements to define the terms.
- Failing to check the baseline may result in attributing changes in the response variable to the setting of the treatment variable when no actual relationship exists and the response variable would have changed regardless of the setting of the treatment variable.
- Blinding may be needed to increase objectivity.

EXAMPLE 9.3

A quality engineer is investigating the root cause of shrinkage porosity in an aluminum die-cast part. The hypothesis is that "shrinkage porosity is the result of insufficient pressure during the casting process." The die-cast machine pressure was set at 150 kN during the production run that resulted in many parts with porosity, so the experiment will determine if higher pressure eliminates the problem. High and low pressures are not operational definitions, so the quality engineer defines low pressure as 150 kN and high pressure as 195 kN. Porosity is defined operationally as an open space on the surface of a die-cast aluminum part.

The treatment variable is casting pressure, and the response variable is the presence or absence of porosity on the finished sample part. Potential confounding variables are material type, material volume, ejector operating cycle, first-stage velocity, and second-stage velocity. The quality engineer records the machine settings for the confounding variables and ensures that they stay the same during each experimental run. The material quantity must also stay the same, and all material used must come from the same source to eliminate variability.

The quality engineer makes a trial run using the previous pressure setting of 150 kN to establish a baseline. If the part produced is without porosity, then the experiment cannot continue without modification; the experimental test condition is expected to produce a part free of porosity. However, the results would be without value if the initial condition were also without porosity. It would not be appropriate to conclude that higher pressure eliminated the porosity and therefore low pressure was the root cause of the porosity.

The baseline experiment resulted in porosity. The experiment then is run under the experimental high-pressure condition, and there is no porosity present. The entire experiment is then repeated five times to ensure the results are consistent.

PROCEDURE

Step 1: Create a test plan based on a hypothesis. The predicted result of the hypothesis is the response variable.
Step 2: Determine the treatment conditions by establishing the treatment variable or variables.
Step 3: Identify potential confounding variables and establish a method to eliminate, control, or minimize them. Blocking and randomization may be useful here.
Step 4: Ensure all terms are written as operational definitions.
Step 5: Establish the baseline if necessary or possible.
Step 6: Perform the experiment.
Step 7: Replicate the experiment and compare the results; a large difference is an indicator that variation is present and more replicates are needed.

10

The Classic Seven Quality Tools

INTRODUCTION

There are many potential tools available to assist in root cause analysis (RCA). The most well-known collection of tools is the seven quality tools that were originally brought together by Kaoru Ishikawa. The tools were intended for RCA performed by production personnel. They are relatively simple to use, yet highly effective as part of an RCA. The classic seven quality tools are an essential part of any RCA tool kit.

ISHIKAWA DIAGRAM

The Ishikawa diagram or cause-and-effect diagram (Figure 10.1), also known as a fish-bone diagram, was created by Kaoru Ishikawa to show the potential causes that could lead to an effect (Ishikawa, 1991). It is also useful to facilitate brainstorming when the potential root cause of a failure is unknown. Factors to be considered are the six *M*s: man (people), measurements, material, milieu (environment), methods, and machines (equipment). These factors can and should be modified to fit an individual process. For example, if the root cause has been localized to a specific area, the factors could be changed to reflect this. However, one should be aware that other factors, such as measurement or material, might still influence the affected process and therefore still be relevant to the root cause.

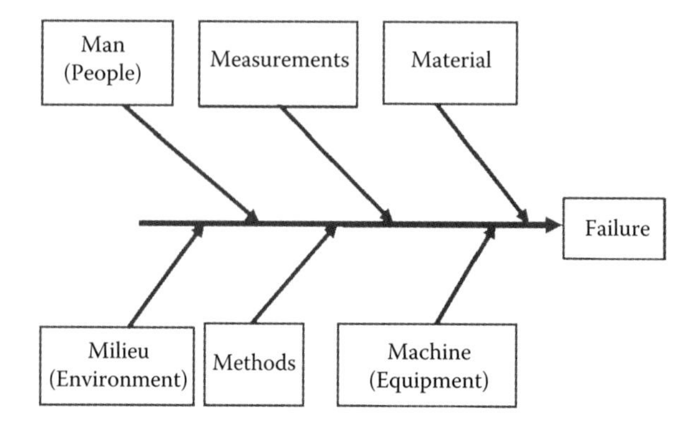

FIGURE 10.1
Ishikawa diagram.

Key Points

- Ideally, these are used with a multifunctional team.
- The factors should at minimum be considered during an RCA.
- Each item in the cause-and-effect diagram should be evaluated. This could mean simply eliminating the item as a possibility or testing to evaluate the contribution of the item to the failure under investigation.
- Each item in the cause-and-effect diagram can be broken down into subitems.

EXAMPLE 10.1
Figure 10.2 provides an example of an Ishikawa diagram.

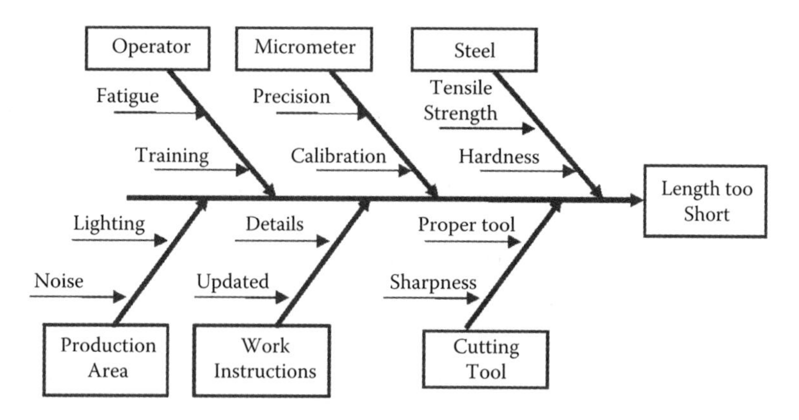

FIGURE 10.2
Ishikawa diagram example.

PROCEDURE

Step 1: Draw a horizontal line with an arrow pointing to the right.
Step 2: List the effect (failure) under investigation in front of the arrow.
Step 3: Draw three arrows below the horizontal line and three arrows above the horizontal line. They should be pointing toward the line and angled toward the right at approximately a 50° angle as per the example in Figure 10.2.
Step 4: Label each of the lines with the six *M*s: man (people), measurements, material, milieu (environment), methods, and machine (equipment). Note: The categories may be modified as needed.
Step 5: Use horizontal arrows pointing toward the angled arrows under each effect category. Here, the individual subcategories of effects should be listed.
Step 6: If necessary, the subcategories can be further broken down with additional lower-level categories.
Step 7: Each potential effect should be evaluated to support or reject it as having an effect on the failure.

CHECK SHEET

Check sheets are used for data collection (Ishikawa, 1991). They are typically used to collect failure occurrence rates. For example, a 100% inspection may use a check sheet to record all inspection results. This information is helpful for determining the actual failure rate. There are different types of check sheets that can be used.

Check Sheet with Tally Marks

A check sheet (Figure 10.3) should have at a minimum three columns: the defect, the number of occurrences, and the total. Care must be used when selecting the defect categories; the defect types should not be so general that multiple types of defects could be entered under the same category, thereby resulting in confusion. The defect categories must also be clear enough so that everybody using the list can assign the defect found to the correct defect category. It may be helpful to have "other" listed as a defect for defects that do not fit into any other category. However, the defects under other should be described to identify them. They may also need their own defect category if they occur too often.

Each time a defect is found, a tally mark is to be entered across from the appropriate defect name and under the heading "number of occurrences."

Defect	Number of Occurrences	Total

FIGURE 10.3
Check sheet example.

Typically, the first through fourth occurrences are listed as individual hash marks, and the fifth is a mark across the first four, such as ~~IIII~~. Each group of five is separated by an empty space to make counting easier.

Tally marks can also be created using dots at the corners of a box to represent one through four and lines connecting the dots to represent five through eight. Nine and ten would be represented by lines crossing the box (Tukey, 1977). Another alternative method to the traditional tally mark is to use a vertical line for the number one, a horizontal line for the number two, a second horizontal line for the number three and a second vertical line for the number four. A diagonal line is used for the number five, and the process is started over for the next group (Figure 10.4). The advantage of this method is that it is faster to add the tally marks and there is less chance of miscounting the groupings.

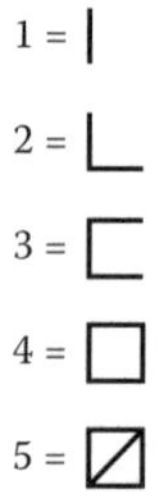

1 =
2 =
3 =
4 =
5 =

FIGURE 10.4
Tally marks.

Defect	Number of Occurrences	Total
Too short	~~||||~~ ~~||||~~ ~~||||~~ //	17
Too long	~~||||~~ /	6
Dented	~~||||~~	5
Other	///	3

FIGURE 10.5
Use of a check sheet example.

Key Points

- Check sheets are used for the collection of failure data.
- The data can be used as the basis for a Pareto diagram.
- The defect categories should be clear enough that different defect types are not inadvertently mixed together and multiple users will constantly assign defects to the appropriate types.

EXAMPLE 10.2
Figure 10.5 shows the use of a check sheet example.

PROCEDURE

Step 1: Make a table containing the following headings: defect, number of occurrences, and total.
Step 2: List the potential defects.
Step 3: Use tally marks to denote each defect item found under each category.
Step 4: Count the groups of tally marks to find the total number of occurrences.

Check Sheet with Graphical Representations

Check sheets with graphical representations are used for data collection during an RCA. It can be effectively used to narrow the actual spatial location of a problem. A floor plan, photo, or other graphical representation is used, and an *X* or other symbol is placed where an item or occurrence of interest is found. A symbol key should be provided if multiple symbols are used, such as when many factors are under investigation.

Key Points

- Check sheets with graphical representations can be used with maps, sketches of an area, process flowcharts, or other graphical representations of an area.
- They can also be used with photos of the item in question.

EXAMPLE 10.3

A hypothetical manufacturing company has a problem with bolts being dropped into a machine on an assembly line. Unfortunately, the bolt size in question is used at multiple workstations, so the exact location of the problem is unknown. Extra bolts are frequently found in both finished product at the final inspection and in the dustpans of the cleaning crew, which indicates that some bolts drop into the product and others fall to the floor.

A quality engineer draws a schematic sketch of the assembly line and performs a twice-daily floor check to find dropped bolts. Each bolt found is indicated by an *X* on the sketch of the production floor (Figure 10.6).

PROCEDURE

Step 1: Create a map, flowchart, photo, or diagram of the entire area under investigation.

Step 2: Determine the time frame for the data collection, for example, every hour or at the end of a shift, day, or week.

Step 3: Check for the factor under investigation.

Step 4: A mark should be placed on the graphical representation where the factor under investigation is found.

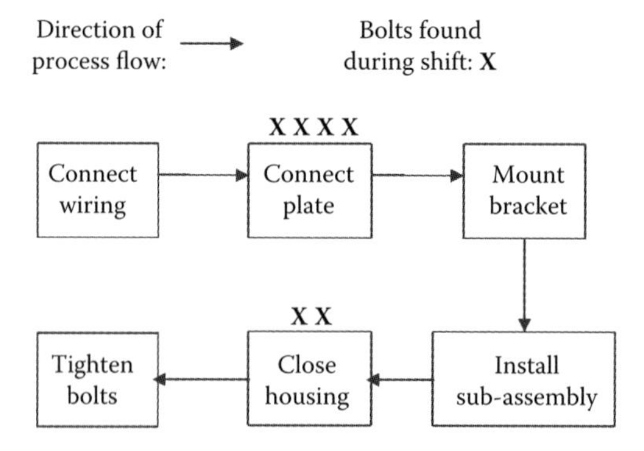

FIGURE 10.6
Check sheet example with floor plan.

RUN CHARTS

Run charts are useful to visualize data points over time (George et al., 2005). They are the basis for statistical process control (SPC); however, a simple run chart does not use specification and control limits like SPC.

The *x* axis is a unit of time, and the *y* axis is the characteristic under consideration. This could be a value such as a measurement or the number of occurrences of a defect or an event. Stratification should be considered. This means separating combined data on the *y* axis. For example, data from multiple production machines or across production shifts can be separated to see if a difference is present.

Key Points

- The *x* axis is the time. This could be a specific date or simply the checks that were made in order of occurrence.
- The *y* axis is either a value such as a measurement or the number of occurrences.
- These data can be used to visualize the occurrences of an issue or the variation within data.
- Stratification is the separation of the data on the *y* axis into layers.

EXAMPLE 10.4

Figure 10.7 provides an example of a run chart.

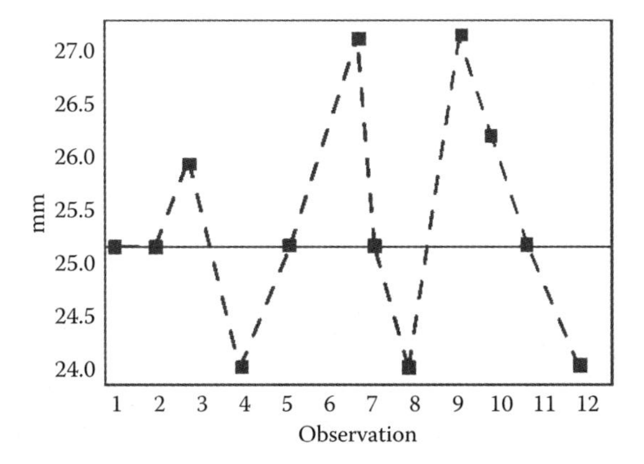

FIGURE 10.7
Run chart example.

PROCEDURE

Step 1: Determine what will be used for the *x* axis.
Step 2: Determine what will be used for the *y* axis.
Step 3: Collect the data.
Step 4: Graph the data.
Step 5: Regraph the data using stratification, if appropriate.

HISTOGRAM

Histograms are useful to visualize the occurrence of events and the distribution of results (Ishikawa, 1991), such as measurement data. The *y* axis of the histogram is the number of occurrences, and the *x* axis is what is being evaluated. Information can be gained by observing the shape of the distribution of data in a histogram; however, appearances may be misleading. The only way to be certain of a distribution is to use a statistical calculation (Figure 10.8). In a normal distribution, data points are concentrated around the mean, and this is the "normal distribution," "bell-shaped distribution," or "Gaussian distribution." A skewed distribution is one in which the data are skewed to one side of the mean. This could be caused by a natural limit such as time, which cannot be less than zero.

Key Points

- The *y* axis is the number of occurrences.
- The *x* axis can be used for either categories such as failure types or measurement data.
- Histograms are used to visualize the distribution of data.

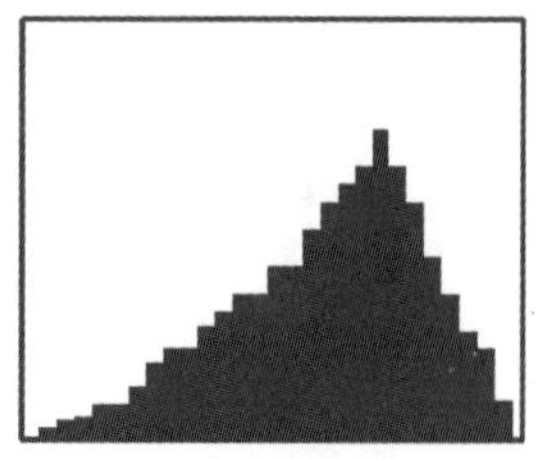

Left-skewed distribution

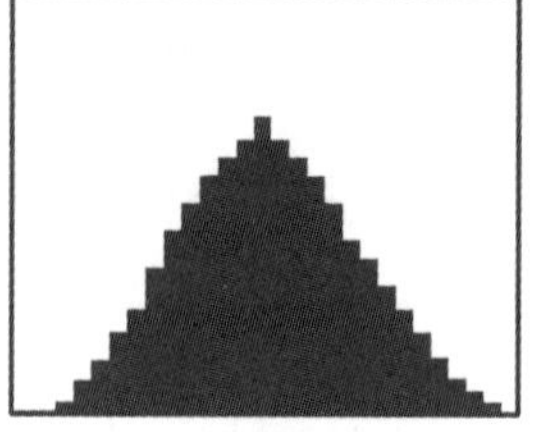

Normal distribution

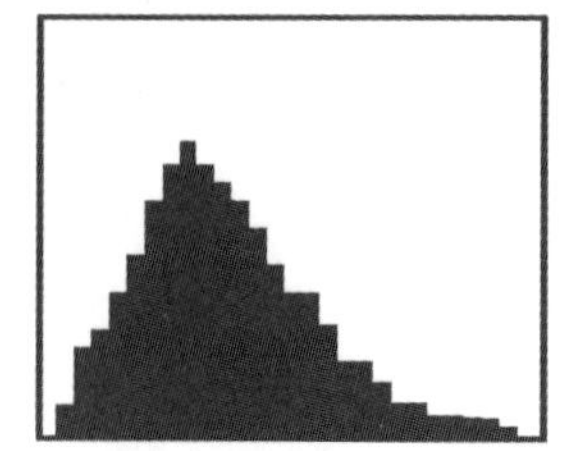

Right-skewed distribution

FIGURE 10.8
Histogram distributions.

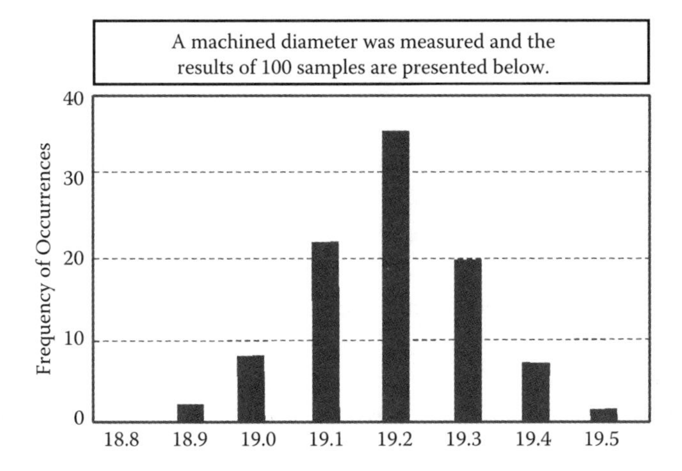

FIGURE 10.9
Histogram example.

EXAMPLE 10.5

A histogram example is given in Figure 10.9.

PROCEDURE

Step 1: Collect the needed data.
Step 2: Determine the number of categories (known as "bins") and list them on the *x* axis.
Step 3: Determine bin size. There is no "best" bin size. Generally, the square root of the total number of data points is used as the bin size.
Step 4: Determine the highest and lowest numbers in the data set. The first bin must include the lowest number, and the last bin must include the highest number.
Step 5: List the frequency of the occurrences on the *y* axis.
Step 6: Use a bar to indicate the number of occurrences for each bin.
Step 7: Observe the pattern of the data and draw conclusions.

PARETO CHART

The Pareto chart is based on the Pareto principle, also known as the 80/20 principle. According to the 80/20 principle, 80% of costs are the result of 20% of the problems (Stockhoff, 1988). The Pareto chart is used to identify the issues for which improvement efforts will deliver the most returns. Note that the Pareto principle should be considered a guide and not a rule; for example, rare safety issues should be addressed before more common issues with less severity.

Key Points

- Identifying the 20% of problems that will result in an improvement of 80% helps to find the optimal prioritization.
- Engineering judgment should be applied when using a Pareto chart; common sense is needed.
- A Pareto chart can be used for failure types, failure locations, failure costs, or other categories as deemed appropriate.
- The main use is determining priorities.

EXAMPLE 10.6

A Pareto table example is provided in Figure 10.10, and Figure 10.11 is a Pareto chart example.

PROCEDURE

Step 1: Determine which categories to use.
Step 2: Gather the relevant data.
Step 3: Create a table with category, total, percentage, and cumulative percentage as the headings.
Step 4: Enter the individual factors under the category column.
Step 5: Determine and enter the total number of occurrences for each factor.
Step 6: Sort the factors so that the highest number of occurrences comes first.
Step 7: Determine the percentage for each factor and enter in the percentage column.
Step 8: Add each percentage in the cumulative percentage column.
Step 9: Create a graph with percentage on the y axis and the factors on the x axis. Use a bar graph to depict the percentage for each category and a line graph to depict the cumulative percentage at each category.
Step 10: Observe the Pareto chart and determine prioritization.

Category	Total	Percentage	Cumulative Percentage
Too short	17	51.5	51.5
Too long	8	24.2	75.7
Dented	5	15.2	90.9
Other	3	9.1	100.0

FIGURE 10.10
Pareto table example.

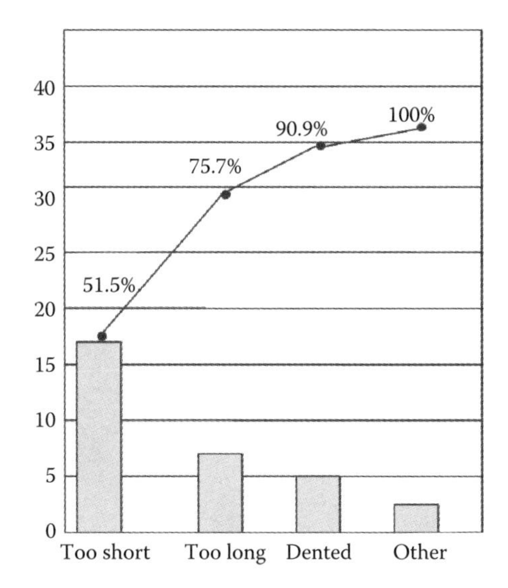

FIGURE 10.11
Pareto chart example.

SCATTER PLOT

Scatter plots, also known as scatter diagrams, are used to analyze possible relationships between paired data (Ishikawa, 1991), such as temperature in producing steel and the resulting hardness of the steel. A scatter plot can indicate strong or weak positive or negative correlations or the absence of a correlation. A correlation does not necessarily mean that the two factors are related; they could be independent of each other and related to a third factor (George et al., 2005).

Key Points

- Scatter plots indicate possible relationships between paired data.
- A correlation may be the result of an unstudied external factor common to both data sets and not the presence of an actual correlation between data sets.

EXAMPLE 10.7
Figure 10.12 provides examples of correlations.

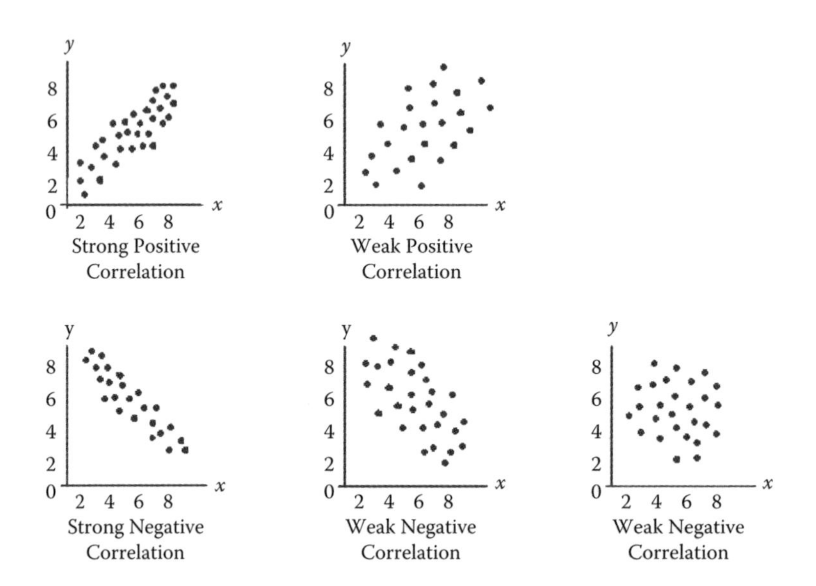

FIGURE 10.12
Examples of correlations.

PROCEDURE

Step 1: Plot each data point on a graph containing an x and a y axis.
Step 2: Observe the data and determine if there is a pattern.

FLOWCHART

Flowcharts are used to graphically depict a process (Brassard and Ritter, 2010). A flowchart can be used to compare the expected operations in a process to the actual operations. It can also be used to locate potential problem areas that should be investigated during RCA. Lines with arrows symbolize the direction of the process. Many symbols for the process steps are possible; the most basic are a box for a process operation and a diamond for decisions.

Key Points

- Flowcharts are helpful for understanding the flow of a process and the multiple possible interactions in the process.
- Care should be used in flowcharting a process as it actually is and not as it is believed to be. Observe the actual process whenever possible.
- Flowcharts can help to identify specific elements of the process for detailed RCA.

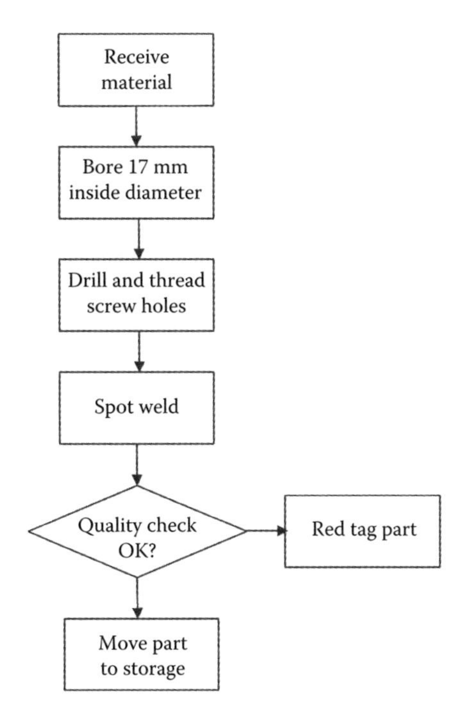

FIGURE 10.13
Flowchart example.

EXAMPLE 10.8
A flowchart example is given in Figure 10.13.

PROCEDURE

Step 1: Enter the first step in the process into a rectangle.
Step 2: Use an arrow to connect the first step to the second step, which is depicted with a box. Repeat for each additional step.
Step 3: Use a diamond for decision points.

11

The Seven Management Tools

INTRODUCTION

The seven management tools are intended for management and planning; however, they can be useful during a root cause analysis. They can help facilitate brainstorming and idea generation sessions because ideas are easier to discuss in large groups when they are written down. These tools are also helpful when deciding on potential improvement actions after the identification of a root cause. Use of these tools is no substitute for an empirical investigation, but they can be a useful supplement.

MATRIX DIAGRAM

A matrix diagram can come in many shapes; however, the most basic type is an L. The concept is simple and easy to use but can be advantageous. There is no limit to the possible uses of matrix diagrams; they can be used for listing what is known about a failure (Figure 11.1), listing roles and responsibilities for teams, comparing multiple factors to each other, or comparing possible improvement options after a root cause is identified. A matrix diagram can also make managing a customer complaint easier by providing a matrix linking supplier and customer part numbers to the company's part number for the failed item.

Key Points

- Many shapes are available; however, the L shape is the most common.
- The matrix diagram serves many purposes, such as listing failures or locations of failures, and can serve as a part number matrix.

(All dimensions are in mm)	Supplier 1 Actual (mm)	Supplier 2 Actual (mm)	Supplier 3 Actual (mm)
Inside diameter 40.0 +/−0.1	Sample 1: 39.97 Sample 2: 39.98	Sample 1: 39.99 Sample 2: 39.99	Sample 1: 40.00 Sample 2: 40.01
Outside diameter 42.0 +/−0.1	Sample 1: 42.05 Sample 2: 42.04	Sample 1: 42.00 Sample 2: 42.00	Sample 1: 42.00 Sample 2: 42.01
Wall thickness 2.0 +/−0.15	Sample 1: 2.08 Sample 2: 2.08	Sample 1: 2.01 Sample 2: 2.01	Sample 1: 2.00 Sample 2: 2.00

FIGURE 11.1
Matrix diagram for a failure analysis.

- It can serve as a tracking list for action items when tasks, due dates, and the people responsible are added.
- The matrix diagram can also be used for presenting multiple options to choose from.
- The data from a matrix diagram can be later used for an is-is not analysis.

EXAMPLE 11.1

A quality engineer is investigating a problem with the diameter of a rolled-steel tube that is provided by three different suppliers. The quality engineer creates an L-shaped matrix and compares the factors inside diameter, outside diameter, and wall thickness on two randomly selected sample parts from each of the three suppliers. Based on the matrix, the quality engineer observes parts from supplier 1 containing more variation and both a larger inside and outside diameter than the other parts. The quality engineer then continues the root cause analysis to determine if this variation is relevant to the problem.

Procedure

Step 1: Determine which factor will be used for the top row. There should be some commonality between the factors in this row, such as machines, shifts, supplier, and part types.

Step 2: Determine what will be compared and list it in the right-hand column.

Step 3: Collect and enter data into the matrix.

ACTIVITY NETWORK DIAGRAM

An activity network diagram is much like a simplified PERT (program evaluation review technique) chart; it is used to identify the critical path, which is the order of operations that can delay a process. The activity network diagram shows which tasks must be started by a specific time to ensure that the entire operation ends on time. This prevents delays and could be useful if process changes are required after a root cause has been identified or for prioritizing activities related to root cause analysis.

Key Points

- The activity network diagram functions like a PERT chart by identifying the tasks or operations that must be completed on time to ensure a project or process finishes on time.
- It is useful for the optimization of a complicated process.
- It can also be used for determining the order to start activities related to root cause analysis.

EXAMPLE 11.2
Figure 11.2 is a PERT chart for process improvement.

PROCEDURE

Step 1: Create boxes for the starting and stopping points.
Step 2: Identify the different processes that must be completed.
Step 3: List the process steps in the order in which they must be completed.
Step 4: Label each process step with the time required for completion.
Step 5: Determine the total time for each process.
Step 6: The process with the greatest time will be the one with the critical path; this process will determine the minimum time to complete all activities.

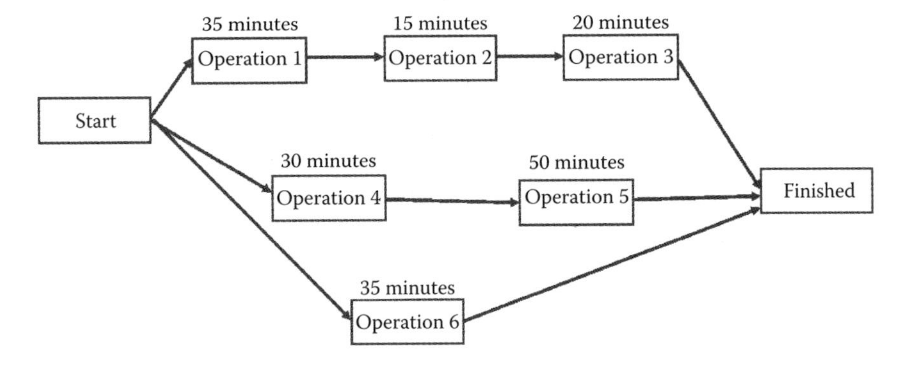

FIGURE 11.2
PERT chart for process improvement.

PRIORITIZATION MATRIX

A prioritization matrix uses weighted values to assess different alternatives, such as possible quality improvements. The alternative under consideration is entered on the left side of the prioritization matrix, and a weighted value is given for every factor. The weighted option should be entered into the column next to the function; the next columns are used to list the options under consideration. The degree to which each option fulfills the requirement needs to be evaluated using numbers that will later be used as multipliers. Here, a large range should be avoided or there may be long discussions, such as deciding if something is a six or a seven. The priority is then multiplied by the multiplier, and each lower assessment in the column is multiplied with the factor and added to this. The total for the column is listed at the bottom of the table; the highest possible evaluation is then determined and used to calculate the percentage that each option fulfilled the weighted values.

Key Points

- A prioritization matrix assists in assessing multiple alternatives.
- It uses weighted values so that aspects that are more important have more weight.
- It can be used for determining which improvement activities should be implemented.

EXAMPLE 11.3

A prioritization matrix for choosing between options is presented in Figure 11.3.

PROCEDURE

Step 1: Create a table.

Step 2: The left column should contain the factors being evaluated.

Step 3: The top of the second column is labeled "Weighted Value."

Step 4: The weighted value for each factor needs to be listed in the weighted value column. A simple weighting system can go from one to three, with three being best. Other weights are possible.

Step 5: Identify the options that will be compared.

Step 6: Assign values to the factors for each option.

Step 7: In the row beneath the last of the factors, write the word *Total* in the weighted value column and write the word *Percentage* beneath this.

	Weighted Value	Option 1	Option 2	Option 3
Ease of use	10	6	1	3
Cost	8	6	1	3
Weight	10	1	6	6
Reliability	9	6	6	3
Service	5	3	6	6
Total:	187	162	171	
Percentage:	74.2	64.3	67.9	
Degree of fulfillment: 1 = not at all, 3 = medium, 6 = 100% fulfillment.				

FIGURE 11.3
Prioritization matrix for choosing between options.

Step 8: Multiply each option's degree of fulfillments multiplier by the weighted value. Add the numbers and write the result in the total row beneath the option.
Step 9: Divide the total by the highest number possible for the weighted values and write the result in the percentage row.
Step 10: Look in the percentage row and determine which option had the highest result. This is the optimal solution according to the given criteria.

INTERRELATIONSHIP DIAGRAM

An interrelationship diagram helps to understand how factors relate to each other. Factors related to the issue under consideration are written on index cards and taped to a flip chart or a wall. The team must fill out the cards, and then they are displayed. The relationships among the factors are identified using arrows.

Interrelationship diagrams can be especially helpful in the investigation of failures when there is little empirical evidence, such as when a service is improperly performed. For example, an order clerk may enter the wrong data when a customer places the order. An interrelationship diagram can display the relationship of all the potential causal factors.

Key Points

- An interrelationship diagram graphically displays the relationship between multiple factors.
- It is useful for the investigation of service failures when there may be multiple influence factors.
- It is easy to create using index cards or a flip chart.

EXAMPLE 11.4

Figure 11.4 provides an interrelationship diagram for quality improvement.

PROCEDURE

Step 1: Write down potential casual factors. Be sure not to cluster the factors too closely together.

Step 2: Use arrows to show the directions of influence between the potential causal factors.

Step 3: Observe the interrelationship diagram and draw conclusions based on the relationships between the factors. Try to identify the factor or factors with the most influence on others. These factors should be considered for improvement actions.

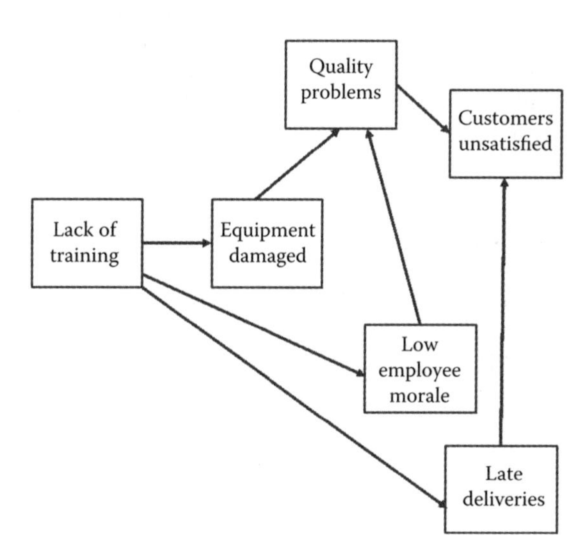

FIGURE 11.4
Interrelationship diagram for quality improvement.

TREE DIAGRAM

A tree diagram has a top level written on top and lower levels are written below. For root cause analysis, the problem investigated is the top category, and possible problems are identified beneath the top category. These are then broken down into lower levels. The concept is much like a fault tree analysis (FTA); however, FTA assigns the probability of occurrence to each category listed. For a new process, these probabilities may be unknown because of lack of data.

Key Points

- A tree diagram is much like an FTA but without assessing the probability of the occurrence of an event or failure.
- It can be used as a failure tree to depict underlying causes and their causes.

EXAMPLE 11.5
A tree diagram for the wrong part shipped is presented in Figure 11.5.

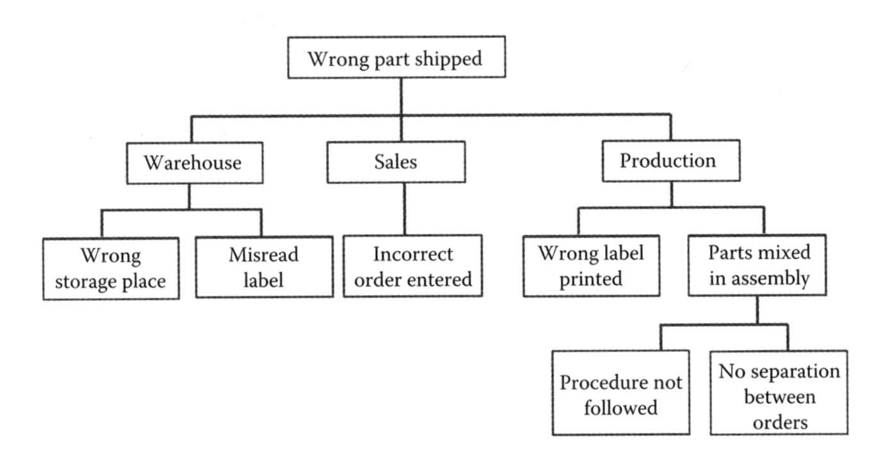

FIGURE 11.5
Tree diagram for wrong part shipped.

PROCEDURE

Step 1: List the failure under investigation.
Step 2: List potential causes or classifications of causes beneath the failure investigated.
Step 3: Connect potential causes to the failure.
Step 4: Repeat for lower-level causes.
Step 5: Analyze the resulting tree diagram and determine which potential causes should be investigated in detail.

PROCESS DECISION TREE CHART

A process decision tree is used to graphically display categories and various levels of details. The main category is identified in a box on the left side, and lower levels are written on the right side. Each level can be broken down into more detail using additional boxes, and a tree diagram can be created on a whiteboard or using index cards. A process decision tree diagram may be helpful when addressing an issue that does not have one clear cause; many potential solutions can be written down and then evaluated.

Key Points

- A process decision tree chart helps when deciding between many possible solutions to a problem.
- It is easy to create and understand.
- It has the same structure as a tree diagram but is horizontal and not vertical like a tree diagram.

EXAMPLE 11.6

Figure 11.6 provides a decision tree for improvement options.

PROCEDURE

Step 1: Write down the subject being considered.
Step 2: List potential options to the right of the subject being considered and connect them to the subject being considered.
Step 3: Additional items can be added to the right of the lower-level items.
Step 4: View the process decision tree and identify the best options.

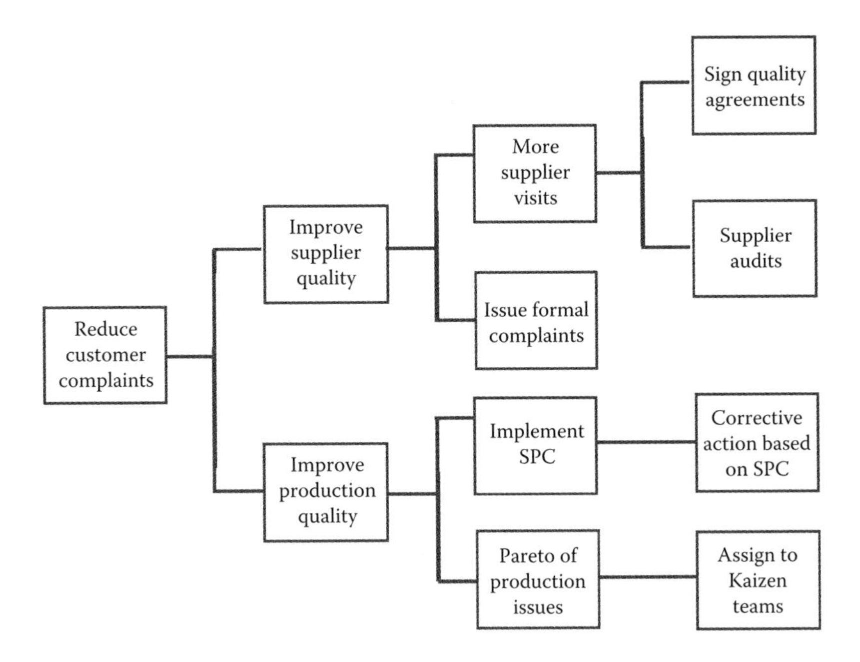

FIGURE 11.6
Process decision tree for improvement options.

AFFINITY DIAGRAM

Affinity diagrams are used to organize ideas based on their relationships to each other. The ideas are first written on index cards; then, the index cards are organized by categories on a whiteboard or wall. Some of the cards may become the top-level category, with others grouped underneath them. This tool can be used to provide a simple structure to a brainstorming session.

Key Points

- The affinity diagram is useful for grouping ideas generated during brainstorming.
- It provides order when confronted with multiple ideas or concepts.
- It can be created using index cards or a dry erase board.
- The affinity diagram requires a team.

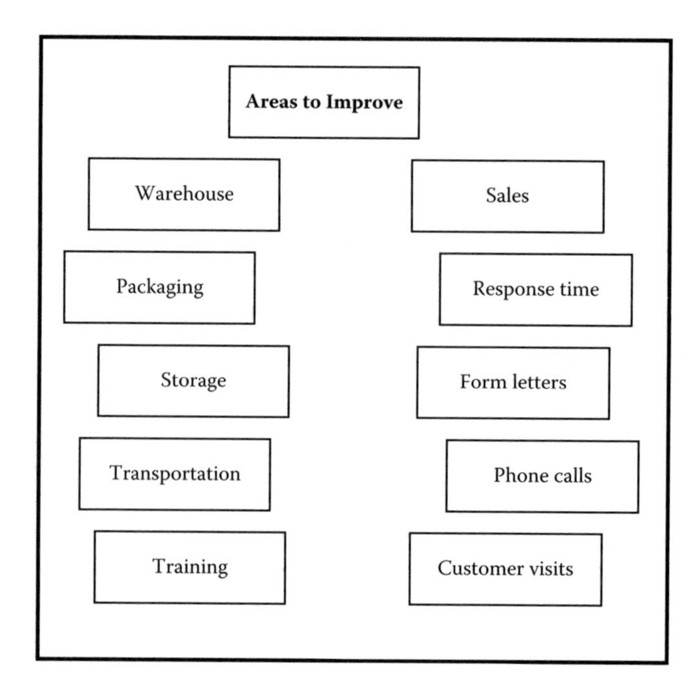

FIGURE 11.7
Affinity diagram for quality improvements.

EXAMPLE 11.7
An affinity diagram for quality improvements is shown in Figure 11.7.

PROCEDURE

Step 1: List all ideas on index cards.
Step 2: Determine top-level categories and write them on index cards.
Step 3: Place the top-level categories on a board.
Step 4: The group must decide which ideas belong under which categories. It may be necessary to vote or duplicate some of the ideas so they can be placed under more than one category.

12

Other Quality Tools for Root Cause Analysis

5 WHY

The 5 Why method is used to dig deeper until the true root cause of an occurrence is identified. The root cause can be pursued by asking why continuously, or the process can be split when an occurrence could have multiple causes (Ohno, 1988). The whys can be, but do not need to be, placed in boxes like a flowchart/fault tree. The 5 Why method should lead from the proximate cause to the ultimate cause.

Key Points

- Asking why five times can lead to the true root cause of an occurrence.
- This method should be used in support of other, more quantitative, methods.
- It can be used for occurrence and failure to detect.

EXAMPLE 12.1

A clamp failed because of wear in the tool used to press it. A quality manager wants to ensure that the true root cause is identified so there will be no recurrence of the issue; the manager creates the 5 Why chart shown in Figure 12.1 to investigate the issue.

PROCEDURE

Step 1: Start with the problem and ask why five times.
Step 2: Each individual why may lead into a new direction of asking why.
Step 3: All whys should be confirmed through other methods.

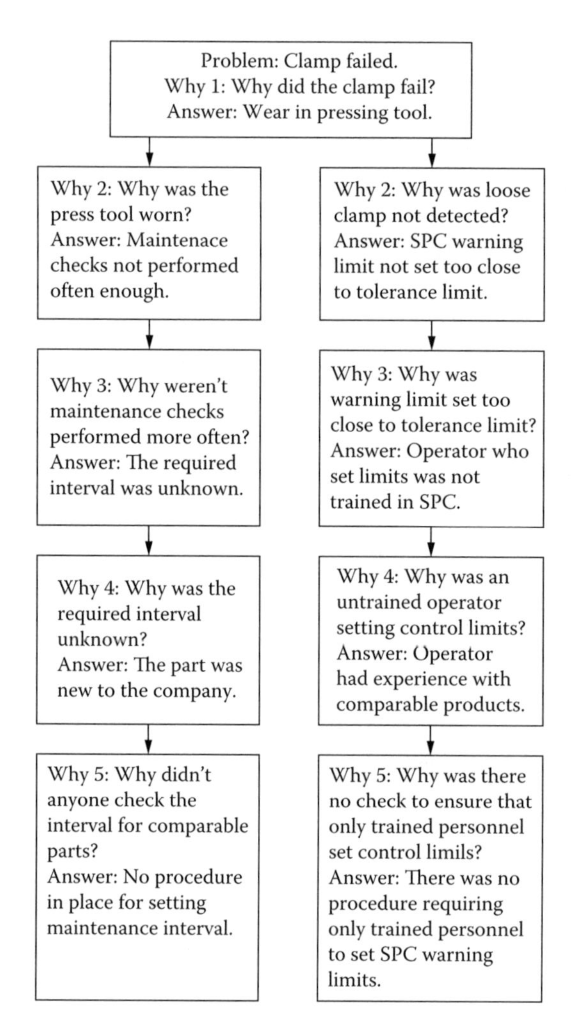

FIGURE 12.1
The 5 Why method for a clamp failure.

CROSS ASSEMBLING

Cross assembling is taking a suspect part out of a failed assembly or system and assembling it into a functioning assembly or system to determine if the problem follows the part. Ideally, the "best- and worst-performing unit" (Bhote, 1991) should be used. This method may lead to insights that should be conformed though other methods. It can also be effective if suspect and known OK parts are compared to determine if a difference exists.

One way to do this is to measure suspect and OK parts. The measurement results can be viewed graphically and compared using statistics to determine if a difference exists.

Key Points

- It is better to use the worst sample if there are varying degrees of nonconformity in a defective part to find a clear difference between OK and suspect parts.
- It is useful for troubleshooting with a system or assembly when the exact defective component is unclear.

EXAMPLE 12.2

A quality engineer is investigating the failure of an actuator assembly. The piston is suspect but does not appear to be out of specification. The suspect piston is removed from its assembly and assembled into a known OK unit to determine if the problem follows it. The problem is now present in the second assembly, so the quality engineer takes the piston from the second assembly and installs it in the first assembly. The unit functions, so the suspect piston is reinstalled in the original actuator. The actuator no longer functions. This indicates to the quality engineer that the piston warrants further investigation.

PROCEDURE

Step 1: Find the worst-possible sample of a suspect part.
Step 2: Assemble the suspect part in an assembly that is known to be OK.
Step 3: Determine if the suspect part still fails to function.
Step 4: Take the corresponding part from the OK assembly and assemble it into the assembly that the suspect part came from.
Step 5: Determine if the OK part also displays the same issue as the suspect part. If so, the problem may be a different part in the assembly and not the suspect part.
Step 6: Return the suspect part to the original assembly and ensure the problem is still present. If the suspect part is a contributing factor, the problem should follow it from assembly to assembly.

IS-IS NOT ANALYSIS

Is-is not analysis is used to help isolate and identify a root cause by determining what could be affected but is not affected by the problem (Tague,

2005). It is useful for eliminating potential root causes and therefore allowing the root cause analysis investigator to concentrate on the potential root causes that may lead to the true root cause.

Key Points

- Factors related to the problem are identified and compared against factors that could be affected by the problem but are not.
- Is-is not analysis is not intended to directly identify the root cause; it is intended to localize the occurrence of the root cause for further investigation.

EXAMPLE 12.3

A quality engineer is unable to determine the cause of sporadic problems with a specific part number. The sample parts checked were in specification, but the end customer is reporting field failures. The quality engineer does not have time to wait for the recall of sample parts and so begins the investigation with only photographs of the part identification tags.

The n.OK parts should come from all machines and shifts but do not. Using the is-is not analysis table presented in Figure 12.2, the quality engineer localizes the potential failure area to machine 2 on the first shift. Looking for more factors that only pertain to machine 2 on the first shift, the quality engineer determines that the parts on this shift and machine all come from a different supplier than the parts that run on other machines. The operators rotate between shifts and machines, and all operators are associated with n.OK parts because of their time on machine 2 during first shift. A detailed investigation of the parts from the suspect supplier leads to the root cause.

Factors	Is	Is not	Difference
Machine	2	1 or 3	Machine 2 not run on shift 2
Shift	1	2	Not all machines run on shift 2
Part number	543-4	542-4	Material type
Material	C1010 steel	High-strength steel	Only part 543-4 uses C1010 steel

FIGURE 12.2
Is-is not for sporadic failures of a part.

PROCEDURE

Step 1: Create a table with the following headings: factors, is, is not, and difference.
Step 2: Determine what factors are involved in the issue under investigation. This should include time, location, and the inputs from an Ishikawa diagram.
Step 3: List the factors in the "factor" column.
Step 4: List what is affected in the "is" column.
Step 5: List what is not affected in the "is not" column. Write "not relevant" if the issue affects everything under this factor.
Step 6: Determine and list the difference between the affected and unaffected factors in the "difference" column. Write "not relevant" if the issue affects everything under this factor.
Step 7: Use this information to narrow down the possibilities in the search for the root cause by determining which factor only affects the factor under investigation.

FOLLOWING LINES OF EVIDENCE

Complex technical investigations seldom have a clear line from point A to point B. Following the lines of evidence can lead to a plausible hypothesis that can account for the observations that have been made. A potential root cause can be identified through multiple lines of evidence, leading to consilience—the point at which multiple hypotheses converge (Wilson, 1999)—and thereby offering more support to the suspicion that the root cause is the true root cause.

Key Points

- Lines of evidence are used to draw conclusions from multiple pieces of evidence.
- They should be used together with other data-gathering tools as well as observations.
- Confirmation testing should be performed to evaluate the resulting hypothesis.
- Using lines of evidence is particularly helpful when there are multiple contradictory lines of evidence.
- Care must be used to differentiate between noise and the root cause.

EXAMPLE 12.4

Approximately 2% of the parts produced by a process are out of specification on length. The same process is used on five machines across three shifts, and the unfinished part comes from two suppliers. Two different types of cutting tools could be used to produce the part.

A quality manager uses other quality tools to better understand the situation, and this investigation determines that the failure happens on all shifts and with all tool types, but only occurs on two of the five possible machines and only involves unfinished parts from one supplier.

These items need to be understood to find the root cause. The supplied parts could be the root cause, but that would not account for the failures only occurring on two of the five production machines. So, the quality manager investigates each item in an attempt to find a hypothesis that could account for all items (Figure 12.3).

Using available quality tools, the quality manager determines that the parts from one supplier are only run on the two machines that are producing the defects. This does not mean that the unfinished parts are the problem, but they warrant further study to determine if they are a contributing factor.

A measurement study of the parts shows that they are all in specification; however, some are on the low side of the tolerance, and the process was not robust enough for these parts. The longer parts are more robust to variation; therefore, they were unaffected. The drawing is changed to tighten the lower tolerance, and the supplier is asked to shift the process mean upward. The defect rate drops to zero.

PROCEDURE

Step 1: List the available evidence in boxes next to each other.

Step 2: Below the boxes, create a box with the word *noise* and a box with the word *root cause*.

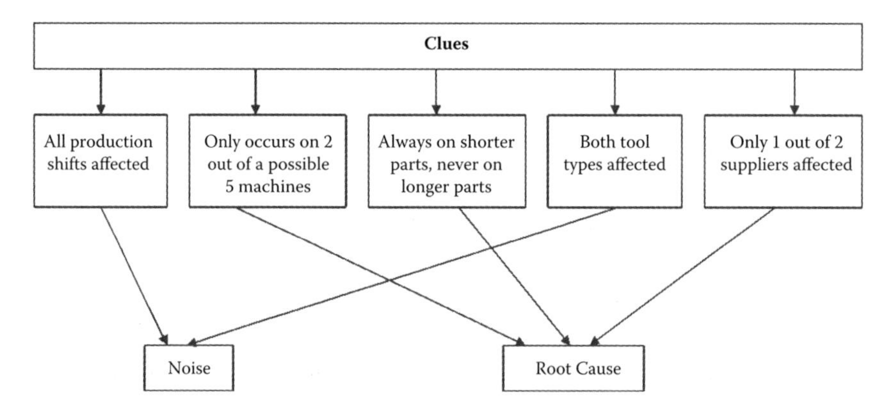

FIGURE 12.3
Lines of evidence for out-of-specification parts.

Step 3: List clues in boxes above the noise and root cause boxes and determine which pieces of evidence fit the root cause.
Step 4: Use arrows to connect the clues to the corresponding root cause.
Step 5: Use arrows to connect any pieces of evidence that do not fit the root cause to the noise box.
Step 6: Form a tentative hypothesis to account for the evidence and evaluate the hypothesis.
Step 7: Repeat the procedure with a new possible root cause if the previous one has been invalidated.
Step 8: Repeat as often as necessary.

PARAMETER DIAGRAM

A parameter diagram (P-diagram) is typically used when designing a product; however, it can be used during root cause analysis to evaluate the system as a whole and to consider factors that could influence the system. This can be especially useful when starting an investigation involving the failure of a complex assembly or analyzing a design failure.

The P-diagram helps to identify parameters that pertain to a system. These parameters are noise factors such as the system's environment, customer usage, interactions with other systems, and variation. Other parameters include input factors, control factors, error states, and the system's ideal function. Use of a P-diagram during root cause analysis can help to identify failures that may have occurred because of noise or a failure in the control factors. A P-diagram is not an empirical method but can identify items for later investigation.

Key Points

- The P-diagram identifies the parameters that affect a system
- It is not a typical root cause analysis tool but can be useful for design failures or when brainstorming.
- Suspect influence factors identified by the P-diagram should be investigated empirically.

EXAMPLE 12.5

A parameter diagram for a bicycle wheel is shown in Figure 12.4.

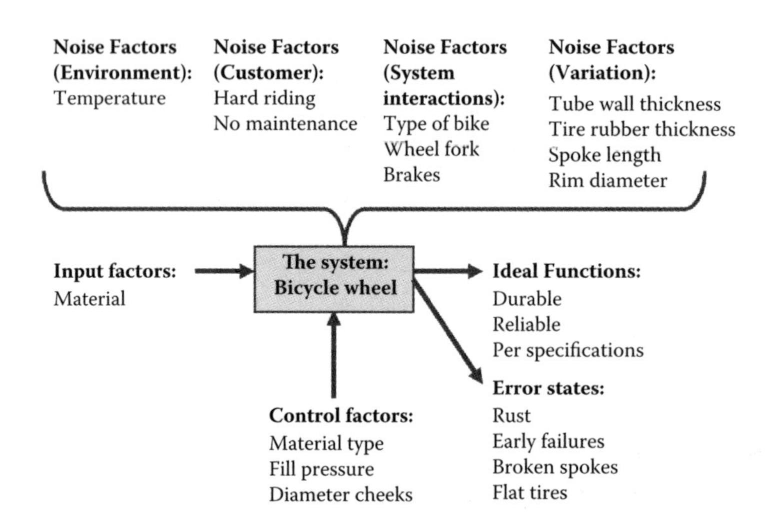

FIGURE 12.4
Parameter diagram for a bicycle wheel.

PROCEDURE

Step 1: Identify the system that will be considered and place it in a box.
Step 2: Identify input factors and place an arrow leading from the input factors to the right toward the system in the block.
Step 3: On the right side of the box, draw an arrow pointing away from the box and identify the ideal functions.
Step 4: Create a second arrow on the right side of the box; this arrow should point down at an angle. Identify the error states here.
Step 5: Draw an arrow pointing toward the bottom of the box and identify control factors here.
Step 6: Identify the types of noise factors that will be used. List the types of noise above the block diagram and use either multiple arrows or a large bracket pointing toward the box. Identify each noise under the appropriate category.

BOUNDARY DIAGRAM

Boundary diagrams are also known as block diagrams because they use blocks to depict the components in a system. This is helpful when determining the limits of a design failure modes and effects analysis (D-FMEA); however, it could also be applied to root cause analysis when analyzing the failure of an assembly within a system. A failing assembly may be in specification and found to be functioning properly when analyzed; the boundary diagram could help to establish other components that could be causing the failure, such as when there is a tolerance stack up involving other components in the customer's system.

An assembly suspected of failing may not be performing properly because of the influence of a different assembly outside the boundary limits, such as when a customer reports a sliding cover on a machine fails to open automatically during the process step when it should be opening. The failure could be caused by mechanical damage in the cover's track, but it may also be caused by an unrelated component. If the cover system is examined and no faults are found, it is possible the sensor that should send the signal to open the cover is the faulty part. The sensor may not be available to analyze if the cover manufacturer did not produce the sensor; in such situations, a boundary diagram can be useful in establishing potential failure causes. In such a situation, it is often necessary to tactfully request that the customer investigate other components in the system.

Key Points

- A boundary diagram is primarily a D-FMEA tool but can be used for root cause analysis.
- It establishes a system's boundary limits.
- It helps to identify outside influences that could cause a system to fail.
- It is sometimes necessary to consider unrelated components that may be causing the system to fail.

EXAMPLE 12.6

Figure 12.5 is a boundary diagram for a water pump.

PROCEDURE

Step 1: Identify the components within the system being considered in the boundary diagram.

Step 2: Create boxes and label them with the components' names. If possible, the boxes should be placed in their relative positions.

Step 3: Draw a dotted line around the system to set the boundary limit.

Step 4: Identify components and assemblies outside the boundary limit.

Step 5: Place the components and assemblies in boxes, ideally in their relative positions.

Step 6: Identify interactions. Possible interactions include mechanical force, heat transfer, fluid transfer, and electricity.

Step 7: Use arrows to show the interactions and the direction of the transfers into or out of the system.

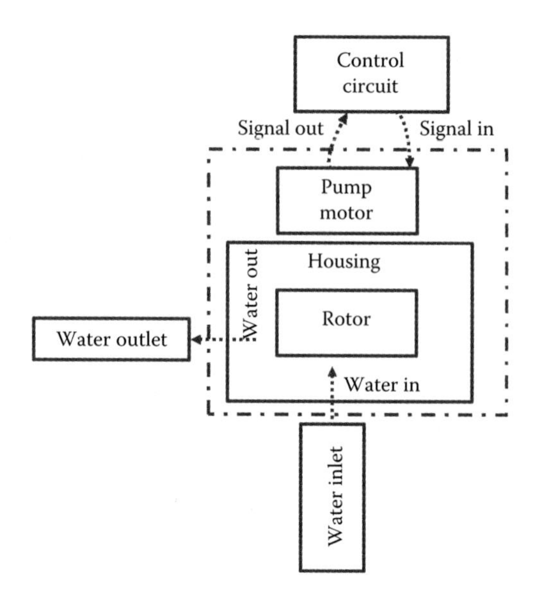

FIGURE 12.5
Boundary diagram for a water pump.

13

Exploratory Data Analysis

INTRODUCTION

Exploratory data analysis (EDA) was created by John Tukey for visually analyzing data as a basis for forming hypotheses and to help determine which methodologies should be used (Tukey, 1977).

Key Points

- EDA is used to form a hypothesis and visualize data.
- Confirmation testing should be performed after EDA has identified a potential root cause.

Procedure

Step 1: Collect data and then graphically display the data using one or more of the EDA methods. Display the data according to time, location, or any other potentially relevant factor.
Step 2: Perform confirmation testing.

STEM-AND-LEAF PLOT

A stem-and-leaf plot is used to graphically depict the distribution of a data set (Montgomery, Runger, and Hubble, 2001). The left side is the stem, and the right side is the leaf. The stem can be used for a single number, such as the tens or hundreds position. The stem can also be used for a range, such as 10–20.

Stem	Leaf
3	23
4	
5	147
6	23579
7	1146
9	268

Stem = tens.

FIGURE 13.1
Stem-and-leaf plot example.

Key Points

- This plot depicts the distribution of the data graphically.
- A stem-and-leaf diagram is much like a histogram.

EXAMPLE 13.1

Figure 13.1 depicts a stem-and-leaf plot for the following numbers: 32, 33, 51, 54, 57, 62, 63, 65, 67, 69, 71, 71, 74, 76, 92, 96, and 98.

PROCEDURE

Step 1: Collect the data.
Step 2: Decide on the unit or range to be used for the stem.
Step 3: Label the columns Stem and Leaf and place a dividing line between them.
Step 4: Arrange the data set in numerical order from lowest to highest.
Step 5: Enter the first part of the first number under the stem column and the remainder of the number under the leaf column.
Step 6: Repeat for all numbers.
Step 7: Observe the stem-and-leaf diagram and look for patterns/tendencies in the data.

BOX-AND-WHISKER PLOTS

Box-and-whisker plots are a method for displaying the distribution of a data set. They can be used when comparing multiple data sets (Tague, 2005), such as the measurement results for parts produced by different machines.

There are many different parts to a box-and-whisker plot. The median (Md) is the point at the middle of the data set. This means half of all data points are above and half are below this point. There is also a box that contains 50% of all data points. The lower side is the first quartile (Q1), and the upper side of the box is the third quartile (Q3).

There are lower inner fences and upper inner fences. Anything below the lower inner fence or above the outer inner fence is an outlier. Below the lower inner fence is a lower outer fence. There is also an upper outer fence above the upper inner fence. Any results beyond the outer fences are extreme outliers. Outliers vary greatly from the rest of the data set. They could be the result of a mistake such as an incorrect measurement or could be the result of a real effect and need additional investigation.

Key Points

- Box-and-whisker plots provide a quick graphical display of the distribution of data.
- They can be used for one data set or to compare multiple data sets.

EXAMPLE 13.2

Figures 13.2 and 13.3 show an example box-and-whisker plot and its details, respectively.

Data set: 290, 335, 353, 354, 354, 355, 356, 356, 361, 363, 365, 367, 368, 369, 378, 385, 405

L (Lowest number in data set) = 290

Q1 = 354

Md (middle of data set or average of two middle points) = 361

Q3 = 368 + 369/2 = 368.5

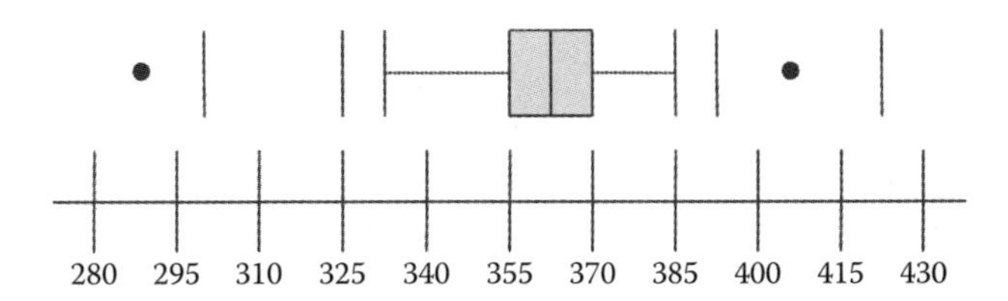

FIGURE 13.2
Example of a box-and-whisker plot.

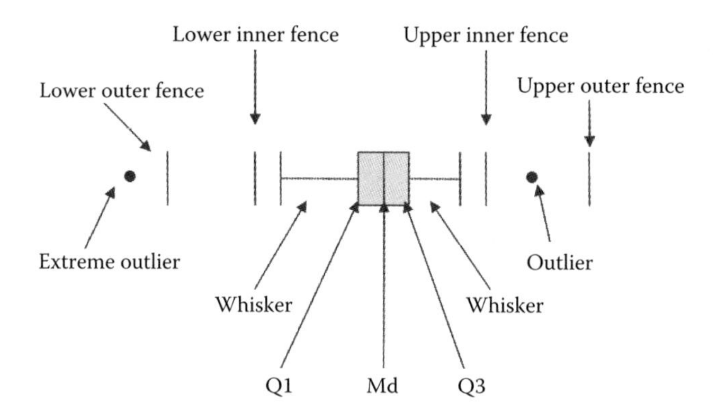

FIGURE 13.3
Details of a box-and-whisker plot.

$$\text{IQR (interquartile range)} = \text{Q3} - \text{Q1} = 368.5 - 350 = 18.5$$

$$\text{H (highest number in data set)} = 405$$

$$\text{Lower inner fence} = \text{Q1} - 1.5 \text{ x IQR} = 354 - (1.5)(18.5) = 326.25$$

$$\text{Lower outer fence} = \text{Lower inner fence} - 1.5 \text{ x IQR} = 326.25 - (1.5)(18.5) = 298.5$$

$$\text{Upper inner fence} = \text{Q3} + 1.5 \text{ x IQR} = 368.5 + (1.5)(18.5) = 396.25$$

$$\text{Upper outer fence} = \text{Upper inner fence} + 1.5 \text{ x IQR} = 396.25 + (1.5)(18.5) = 424$$

PROCEDURE

Step 1: Collect data.
Step 2: Arrange the numbers in order.
Step 3: Place the numbers on a number line.
Step 4: Determine the Md. This will be the number in the middle of the data set or the average of the two middle numbers if there is an even number of numbers. Place a vertical line above the Md on the number line.
Step 5: Determine the Md for Q1. The Md for Q1 is the middle point between the lowest number and the number prior to the Md of the data set. Place a vertical line above Q1 on the number line.
Step 6: Determine the Md for Q3. The Md for Q3 is the middle point between the lowest number after the Md of the data set and the highest number. Place a vertical line above Q3 on the number line.
Step 7: Connect the three vertical lines with a horizontal line above and below. This will form a box containing 50% of the range of the data set.

Step 8: Determine the IQR by subtracting Q1 from Q3.
Step 9: Determine the lower inner fence by subtracting 1.5 x IQR from Q1. Draw a vertical bar above this point on the number line.
Step 10: Determine the upper inner fence by adding 1.5 x IQR to Q3. Draw a vertical bar above this point on the number line.
Step 11: Determine the lower outer fence by subtracting 1.5 x IQR from the lower inner fence. Draw a vertical bar above this point on the number line.
Step 12: Determine the upper outer fence by adding 1.5 x IQR to the upper inner fence. Draw a vertical bar above this point on the number line.
Step 13: Add the first whisker by drawing a dotted line from Q1 to the last data point before the lower inner fence.
Step 14: Add the second whisker by drawing a dotted line from Q3 to the last data point before the upper inner fence.
Step 15: Any data point between the inner and outer fence is an outlier and should be identified with a small circle.
Step 16: Any data point below the lower outer fence or above the upper outer fence is an extreme outlier and should be identified with a circle.
Step 17: Attempt to determine the reason for outliers; for example, they could be a mistake in data collection or measurement error and not relevant or may provide valuable information regarding the factor under investigation.
Step 18: Observe the spread of the data and draw tentative conclusions.

MULTI-VARI CHART

Multi-vari charts are used for displaying and analyzing variation-related data for multiple factors within a single chart (Sheehy et al., 2002). The data that are analyzed can be temporal, positional, or cyclical. Temporal data are data collected from different times, positional data are taken from different locations, and cyclical data are data are taken from regularly scheduled events.

Key Points

- Multi-vari charts depict the variance of factors.
- They can be based on existing data or planned studies.
- A sampling plan needs to be created either to fit the preexisting data or for the collection of the required data.
- Multiple multi-vari charts can be compared to each other by placing them side by side.

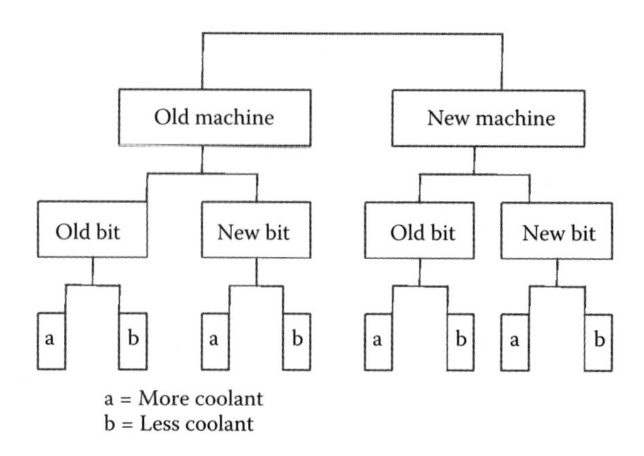

FIGURE 13.4
Example of a multi-vari chart sampling plan.

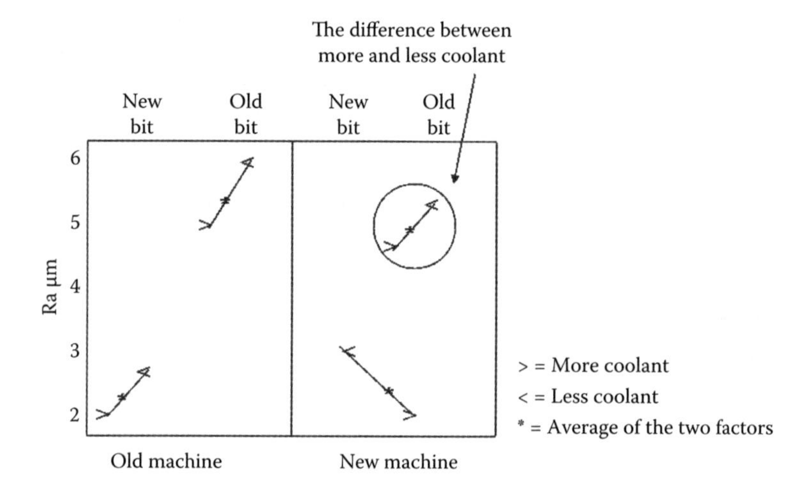

FIGURE 13.5
Example of multi-vari chart results for coolant.

EXAMPLE 10.3

Figure 13.4 is an example of a multi-vari chart sampling plan. Figures 13.5 through 13.7 show multi-vari chart results for coolant, machine, and bit, respectively.

PROCEDURE

Step 1: Start a sampling plan by determining which three factors will be considered.
Step 2: Determine the levels of each factor.

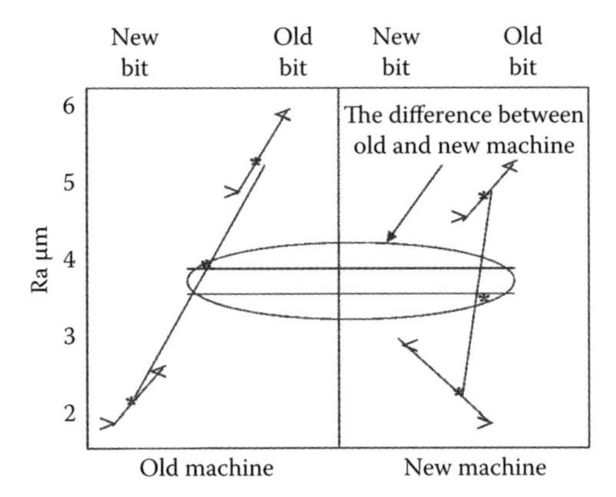

FIGURE 13.6
Example of multi-vari chart results for machine.

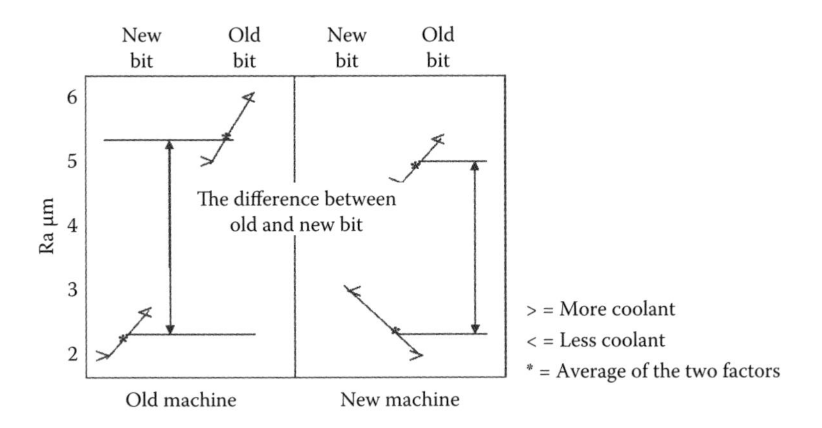

FIGURE 13.7
Example of multi-vari chart results for bit.

Step 3: Perform experiments or gather the relevant preexisting data. Multiple measurement results will be necessary for each factor.
Step 4: Graph the results by placing the output under consideration on the *y* axis.
Step 5: Divide the graph into the separate fields for the first factor. Label the fields below the graph.
Step 6: Label the top of each field with the second factor.
Step 7: Define symbols to represent the highest and lowest results for the third factor and use a symbol to identify the average of the results.
Step 8: Plot the results in the graph.
Step 9: The difference between the averages for each factor is the variation resulting from the changes in each factor.

14

Customer Quality Issues

PLAN-DO-CHECK-ACT FOR IMMEDIATE ACTIONS

The PDCA (**Plan-Do-Check-Act**) cycle can be used for immediate actions when a problem is detected (Figure 14.1). The method of addressing the problem needs to be determined. This could mean deciding to use an 8D report or some other problem-solving approach or addressing the problem with a small problem-solving team. The members of the team should be selected based on the support needed. For example, logistics and salespeople may be needed to determine why the wrong part was shipped to a customer. A qualified person needs to decide if a containment action is necessary as well as the type of containment. A containment action can be as simple as inspecting parts in inventory or as complex as a recall if parts that could affect safety are at the end customer. The containment action then needs to be implemented. If sufficient resources are available, the containment action and root cause analysis can happen in parallel.

Key Points

- An approach for solving the problem needs to be identified.
- An interdisciplinary team is needed.
- The type of resources and support required must be identified.
- Not every situation requires a containment action; however, a decision must be made regarding whether a containment action is needed.

EXAMPLE 14.1

The operator of a production machine discovers a defective component in a batch of material that had been produced during a previous manufacturing process. The machine operator informs a quality engineer, who quickly calls together the supervisor of the two manufacturing departments; the

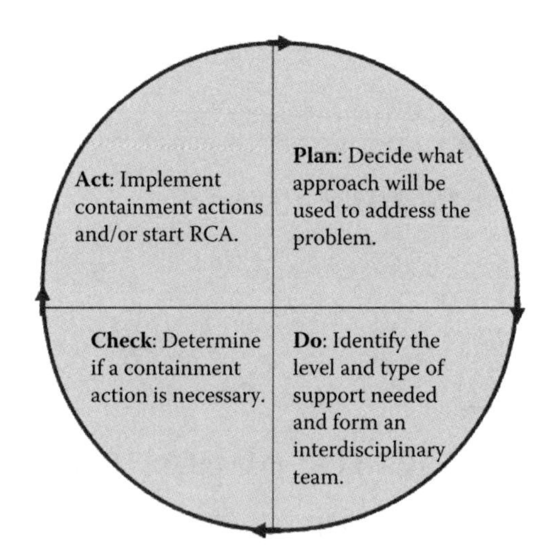

FIGURE 14.1
PDCA for immediate actions.

three decide that it is possible for a defective part to go undetected, so a containment action is needed. The quality engineer informs the warehouse to quarantine all parts, and the supervisor of the department that made the mistake sends three people to the warehouse to begin checking parts. A root cause analysis is started by the quality engineer while the inspection is taking place.

PROCEDURE

Step 1: Determine how the problem will be addressed.
Step 2: Form an interdisciplinary team based on the support needed to investigate the issue and solve the problem.
Step 3: Decide if containment actions are necessary and what type of actions should be taken if containment is necessary.
Step 4: If required, the containment actions need to be implemented.
Step 5: The root cause analysis should start.

8D REPORT

An 8D report is used for reporting on a quality failure. This failure can be internal, such as when one department detects a failure from another department, or external, such as when a customer issues a quality complaint. The 8D report will not identify the root cause of a failure, but it will require a team approach (Palady and Snabb, 2000) as well as containment and improvement actions in addition to the root cause analysis. The 8D report provides a simple and concise form for reporting on a root cause analysis as well as a method of tracking activities.

Key Points

- The 8D report provides a brief, but detailed, report on root cause analysis activities as well as the root cause.
- It can be used for internal or external failures.
- An 8D report is typically used for the investigations of failures, but the same format can be applied to improvement activities.
- Different companies may use different names for the eight steps; however, the steps themselves are the same.
- The fields on top of an 8D report should be customized to fit the needs of the company using the 8D report.

EXAMPLE 14.2

A customer issues a quality complaint because two units were found to be missing bolts. The supplier initiates an 8D report to document the activities pertaining to the failure (Figure 14.2). The inventory was checked, and the use of quality tools led the team to conclude that a new worker failed to install the bolts. Although this was a human error, an automatic check was installed in the process to ensure an assembly with missing bolts could not be shipped to the customer again. The work instruction, process failure modes and effects analysis (P-FMEA) and control plan were found to lack this point, so they were updated.

PROCEDURE

Step 1: Start the 8D report by filling in the fields at the top of the document and forming an interdisciplinary team.
Step 2: Describe the problem.
Step 3: Decide if containment actions are necessary. If so, assign somebody to implement them and report on the results of the containment actions.
Step 4: Identify the root cause of the failure by investigating the part or process that failed. The investigation should be supported by the use of quality tools.
Step 5: After the identification of the root cause, the corrective actions should be described. Trials should be performed to ensure that these actions will be effective.
Step 6: The implemented corrective actions need to be described.
Step 7: Actions must be taken to prevent a reoccurrence of the issue.
Step 8: The team needs to be congratulated, and the report should be closed.

<table>
<tr><td rowspan="2">Report No.: 2014–008
Customer Claim No. 20047595
Supplier 8D No.: 8774</td><td colspan="2">Part No.: 95841-a</td></tr>
<tr><td colspan="2">Complaint: Bolt missing on assy.</td></tr>
<tr><td>(1) Team J. Smith-Team Leader
B. Jones-Production
C. Raynolds-Chamion</td><td colspan="2">Opened on: 18 July 2014
Version date: 4 August 2014</td></tr>
<tr><td colspan="3">(2) Problem Description
Customer found two assemblies in their production with bolts missing.</td></tr>
<tr><td colspan="2">(3) Immediate Containment Action
Customer is inspecting their inventory. (800 parts)
100% check of parts in warehouse. (1,200 parts)</td><td>Resp./Date
Cust. 18 Jul. 14
J. Smith 18 Jul. 14</td></tr>
<tr><td colspan="3">(4) Root Cause
New employee forgot attachment of bolts. This step was not in the work instruction and no additional checks were performed.</td></tr>
<tr><td colspan="2">(5) Planned Corrective Actions
Updating work instructions.
Addition of a go/no-go gage in the process to ensure presence of bolts.</td><td>Resp./Date
B. Jones 23 Jul. 14
B. Jones 1 Aug. 14</td></tr>
<tr><td colspan="2">(6) Implemented Corrective Actions
Work instruction updated.
Go/no-go gage installed and verified.</td><td>Resp./Date
B. Jones 23 Jul. 14
B. Jones 1 Aug. 14</td></tr>
<tr><td colspan="2">(7) Actions to Prevent Reoccurrence

P-FMEAP-FMEA 8921b updated.
D-FMEA
Control Plan Control plan 478c updated.
Procedures Work Instruction 72a updated.</td><td>Resp./Date

B. Jones 23 Jul. 14

B. Jones 23 Jul. 14
B. Jones 23 Jul. 14</td></tr>
<tr><td colspan="2">(8) Congratulate the Team
Thanks for a job well done!</td><td>Closed on
4 August 2014
By J. Smith</td></tr>
</table>

FIGURE 14.2
An 8D report for a customer complaint.

CORRECTIVE ACTIONS

Corrective actions are necessary after identification and confirmation of the root cause of a failure. The corrective actions should be sufficient to ensure that the failure cannot occur again and should be observed to ensure that they have long-term effectiveness. The corrective actions may also be advantageous for other parts or processes for which the failure has not occurred but could be expected to occur. The PDCA cycle should be used to guide the implementation of corrective actions.

Key Points

- Use the PDCA cycle for corrective actions.
- Ensure the corrective actions function long term.
- Look for opportunities to implement the corrective actions as preventive actions for other parts or processes.

EXAMPLE 14.3

A quality manager determined the root cause of a failure was free play in an adjustment knob on a production machine. The corrective actions were an immediate removal of the free play and a check for free play was entered into the maintenance plan for the machine. This change eliminated the problem, so other machines were checked for comparable adjustment knobs. Three machines were found to have comparable adjustment knobs, and these also received updated maintenance plans as a preventive action.

PROCEDURE

Step 1: Identify a corrective action.
Step 2: Implement the corrective action.
Step 3: Check the long-term effectiveness of the corrective action.
Step 4: Look for opportunities to implement the corrective action.

References

Beekman, Christopher S. and Alexander F. Christensen. Controlling for Doubt and Uncertainty through Multiple Lines of Evidence: A New Look at the Mesoamerican Nahua Migrations. *Journal of Archaeological Method and Theory* 10 No. 2 (June 2003): 111–164.

Benbow, Donald W. and T.M. Kubiak. *The Certified Six Sigma Black Belt Handbook*. Milwaukee, WI: ASQ Quality Press, 2009.

Bhote, Keki R. *World Class Quality: Using Design of Experiments to Make it Happen*. New York: AMACON, 1991.

Bisgaard, Soren. *The Role of Scientific Problem Solving and Statistics in Quality Improvement: Some Perspectives*. Center for Quality and Productivity Improvement Report Number 158. Madison, WI: Center for Quality and Productivity Improvement, March 1997. Accessed November 22, 2012, from http://cqpi.engr.wisc.edu/system/files/r158.pdf.

Blackburn, Simon. *Oxford Dictionary of Philosophy*. 2nd ed. Oxford, UK: University Press, 2005.

Borrer, Connie, M. (ed.). *The Certified Quality Engineer Handbook*. 3rd ed. Milwaukee, WI: ASQ, Quality Press, 2009.

Box, George E.P. Science and Statistics. *Journal of the American Statistical Association* 71 No. 356 (December 1976): 791–799.

Box, George E.P. *Statistic for Discovery*. Center for Quality and Productivity Improvement Report Number 179. Madison, WI: Center for Quality and Productivity Improvement, March 2000. Accessed September 29, 2012, from http://cqpi.engr.wisc.edu/system/files/r179.pdf.

Box, George E.P., Stuart Hunter, and William G. Hunter. *Statistics for Experimenters: An Introduction to Design, Data Analysis and Model Building*. 2nd ed. Hoboken, NJ: Wiley, 2005.

Brassard, Michael. *The Memory Plus: Tools for Continuous Improvement and Effective Planning* (Revised ed.). Salem, NH: GOAL/QPC, 1996.

Brassard, Michael and Diane Ritter. *The Memory Jogger 2: Tools for Continuous Improvement and Effective Planning*. 2nd ed. Salem, NH: GOAL/QPC, 2010.

Breyfogle, Forrest W., III. *Implementing Six Sigma: Smarter Solutions Using Statistical Methods*. 2nd ed. Hoboken, NJ: Wiley, 2003.

Chrysler, Ford, General Motors Supplier Quality Requirements Task Force. *Potential Failure Modes and Effects Analysis: Reference Manual* (4th ed.). USA: Automotive Industry Action Group, 2008.

Creswell, John W. *Research Design: Qualitative, Quantitative, and Mixed Methods Approaches*. 2nd ed. Sage Publications: London, 2003.

de Groot, Adriaan D. *Methodology: Foundations of Inference and Research in the Behavioral Sciences*. Translated by J.A.A. Spiekerman. The Hague: Mouton, 1969.

Dekker, Sidney. *Drift into Failure: From Hunting Broken Components to Understanding Complex Systems*. Surrey, England: Ashgate, 2011.

Del Vecchio, R.J. *Understanding Design of Experiments: A Primer for Technologists*. Cincinnati, OH: Hanser, 1997.

de Mast, Jeroen and Benjamin P.H. Kemper. Principles of Exploratory Data Analysis in Problem Solving: What Can We Learn from a Well-Known Case? *Quality Engineering* 21 No. 4 (2009): 366–375.

de Mast, Jeroen and Albert Trip. Quality Improvement Projects. *Journal of Quality Technology* 39 No.4 (2007): 301–311.

Deming, W. Edwards. *Out of the Crisis*. Cambridge, MA: Massachusetts Institute of Technology, 1989.

Deming, W. Edwards. *The New Economics: For Industry, Government, Education*. 2nd ed. Cambridge, MA: MIT Press, 1994.

Department of Defense. *MIL-STD-1520C: Corrective Action and Disposition System for Nonconforming Material*. Washington, DC: Department of Defense, 1986.

Dias, Sónia and Pedro Manuel Saraiva. Use Basic Quality Tools to Manage Your Processes. *Quality Progress* 37 No. 8 (August 2004): 47–63.

Doggett, Mark A. Root Cause Analysis: A Framework for Tool Selection. *Quality Management Journal* 12 No. 4 (2005): 34–45.

Doyle, Arthur Conan. A Scandal in Bohemia. In *The Adventures of Sherlock Holmes*. London: Penguin Books, 1994. pp. 3–29.

Duffy, Grace, Scott A. Laman, Pradip Mehta, Govind Ramu, Natalia Scriabina, and Keith Wagoner. Beyond the Basics. *Quality Progress* 45 No. 6 (April 2012): 18–29.

Feynman, Richard, P. *The Meaning of It All: Thoughts of a Citizen-Scientist*. Reading, MA: Perseus Books, 1988.

Gabor, Andrea. *The Man Who Discovered Quality: How W. Edwards Deming Brought the Quality Revolution to America—The Stories of Ford, Xerox, and GM*. New York: Random House, 1990.

George, Michael L., David Rowlands, Mark Price, and John Maxey. *The Lean Six Sigma Pocket Tool Book*. New York: McGraw-Hill, 2005.

Griffith, Gary K. *The Quality Technician's Handbook*. 5th ed. Upper Saddle River, NJ: Prentice Hall, 2003.

Gryna, Frank M. *Quality Planning and Analysis*. 4th ed. New York: McGraw-Hill, 2001.

Hare, Lynne B. Statistics Roundtable: Chapter One. *Quality Progress* 35 No. 8 (August 2002): 77–79.

Hein, Morris and Susan Arena. *Foundations of College Chemistry*. 10th ed. Pacific Grove, CA: Brooks/Cole, 2000.

Imai, Maaaki. *Kaizen: The Key to Japan's Competitive Success*. New York: McGraw-Hill, 1986.

Imai, Maaaki. *Gemba Kaizen: A Commonsense, Low-Cost Approach to Management*. New York: McGraw Hill, 1997.

Ishikawa, Kaoru. *What Is Total Quality Control: the Japanese Way*. Translated by David J. Lu. London: Prentice Hall, 1985.

Ishikawa, Kaoru. *Guide to Quality Control*. 2nd ed. Translated by Asian Productivity Organization. Tokyo: Asian Productivity Organization, 1991.

Juran, J.M. *Managerial Breakthrough*. New York: McGraw-Hill, 1995.

Juran, J.M. The Non-Pareto Principle—Mea Culpa. In Stephens, K.S. (ed.), *Juran, Quality and a Century of Improvement*. Milwaukee, WI: American Society for Quality, Quality Press, 2005, pp. 185–189.

Kolesar, Peter J. What Deming Told the Japanese in 1950. *Quality Management Journal* 2 No. 1 (September 1994): 9–24.

Levesque, Justin and H. Fred Walker. The Innovation and Process Quality Tools. *Quality Progress* 40 No. 7 (July 2007): 18–22.

Liu, Shu. Tool Time: Seven New Quality Tools Aid Appreciative Inquiry. *Quality Progress* 46 No. 4 (April 2013): 30–36.

Medawar, P.B. *The Threat and the Glory: Reflections on Science and Scientists.* New York: Harper Collins, 1990.

Montgomery, Douglas C. *Design and Analysis of Experiments.* 4th ed. New York: Wiley, 1997.

Montgomery, Douglas C., George C. Runger, and Norma F Hubble. *Engineering Statistics.* 2nd ed. New York: Wiley, 2001.

Ohno, Taiichi. *Toyota Production System: Beyond Large-Scale Production.* Translation by Productivity. Portland, OR: Productivity Press, 1988.

Palady, Paul and Thomas Snabb. *TAPS: A Total Approach to Problem Solving.* USA: PAL, 2000.

Platt, John R. Strong Inference. *Science,* 146, No. 3642 (1964): 347–353.

Popper, Karl. *The Logic of Scientific Discovery*. London: Routledge, 2007.

Quine, W.V. and J.S. Ullian. *The Web of Belief.* 10th ed. New York: Random House, 1978.

Rambaud, Laurie. 8D Structured Problem Solving; A Guide to Creating High Quality 8D Reports. Breckenridge, CO: PHRED Solutions, 2011.

ReVelle, Jack. Keep Your Toolbox Full. *Quality Progress* 45 No. 2 (February 2012): 48–49.

Rooney, James J., T.M. Kubiak, Russ Westcott, R. Dan Reid, Keith Wagoner, Peter E. Pylipow, and Paul Plsek. Building from the Basics. *Quality Progress* 42 No. 1(January 2009): 18–29.

Russell, Bertrand. *The Problems of Philosophy*. Mineola, NY: Dover, 1999.

Sagan, Carl. *The Demon Haunted World: Science as a Candle in the Dark.* New York: Ballantine Books, 1996.

Sandholm, Lennart. Global Influence—A True Mentorship in Retrospect. In Stephens, K.S. (ed.), *Juran, Quality and a Century of Improvement.* Milwaukee, WI: American Society for Quality, Quality Press, 2005, pp. 66–74.

Sheehy, Paul, Daniel Navarro, Robert Silvers, and Victoria Keyes. *The Black Belt Memory Jogger: A Desktop Guide for Six Sigma Success*. Salem, NH: GOAL/QPC, 2002.

Soderborg, Nathan R. Design for Six Sigma. *Six Sigma Forum Magazine* 4 No. 1 (2004): 15–22.

Stockhoff, Brian, A. Core Tools to Design Control, and Implement Performance. In Juran, J.M., and Joseph A. de Feo (eds.), *Juran's Quality Control Handbook.* 4th ed. New York: McGraw-Hill, 1988, pp. 891–941.

Tague, Nancy R. *The Quality Toolbox.* 2nd ed. Milwaukee, WI: ASQ Quality Press, 2005.

Tramel, Stephen. Explanatory Hypotheses and the Scientific Method. In Fort Hayes State University (ed.), *Ways of Knowing in Comparative Perspective: The WKCP Companion and Anthology*. Aceton, MA: Copley Custom Textbooks, 2006, pp. 21–26.

Tukey, John W. *Exploratory Data Analysis*. Reading, MA: Addison-Wesley, 1977.

Valiela, Ivan. *Doing Science: Design, Analysis, and Communication of Scientific Research.* New York: Oxford University Press, 2009.

Wade, Carole and Carol Travis. *Psychology.* 4th ed. New York: Harper Collins, 1996.

Wilson, Edward O. *Consilience*. New York: Random House, 1999.

Wilson, Paul F., Larry D. Dell, and Gaylord F. Anderson. *Root Cause Analysis: A Tool for Total Quality Management*. Milwaukee, WI: ASQ Quality Press, 1993.

Appendix 1: Using the Tool Kit

Tool selection is an important element of root cause analysis. Using the improper tool could waste time and delay the discovery of the root cause. Also important is the ability to quickly switch to a different tool at the appropriate time. Figure A.1 is a chart to help guide tool selection.

Key: Hypothesis generation: ■ Relationship analysis: ● Process analysis: — Data collection: ‡ Decision making: +	
Tool	**Use**
Ishikawa Diagram	—■
Check Sheet with Tally Marks	‡■
Check Sheet with Graphical Representations	‡■
Run Charts	‡—■
Histogram	■
Pareto Chart	■
Scatter Plot	●
Flow Chart	—■
5 Why	●
Cross Assembling	●
Is-Is Not Analysis	●
Following Lines of Evidence	●
Stem-and-Leaf Plot	●
Box-and-Whisker Plot	●
Multi-Vari Chart	●
Matrix Diagram	●
Activity Network Diagram	●
Prioritization Matrix	+
Interrelationship Diagram	■●
Tree Diagram	■
Process Decision Tree Chart	■●
Affinity Diagram	●
Parameter Diagram	■●
Boundary Diagram	■●

FIGURE A.1
Tool selection matrix.

Appendix 2: Terminology

accuracy: The closeness of a measurement result to the true value of what is being measured. Example: A measurement of 5.01 mm is more accurate than a measurement of 5.02 mm if the true value of a measured block is 5.00 mm long.

baseline: The initial condition or data describing the initial condition of a process prior to an experiment. A baseline is used for comparisons.

blinding: Preventing an experimenter from knowing data about an experiment that could result in a conscious or subconscious bias. Example: A technician pushes a button on a digital micrometer to electronically collect the results without looking at the display and possibly inadvertently applying more or less pressure and unconsciously influencing the results.

blocking: The mixing of confounding variables across all experimental sets to minimize the impact of the confounding variables on the results and to reduce variability and thereby increase precision. Example: A quality engineer wants to perform an experiment on parts produced by several machines without having the results unduly influenced by the machines, so he ensures that each sample group includes an equal number of parts from each machine.

confounding: The effects of uncontrolled variables or noise being mixed into the results of an experiment. Example: An experiment using samples produced under different conditions will experience confounding if variation is present and not accounted for.

confounding variable: The variable or factor in an experiment that may be hidden or uncontrollable. Also referred to as noise. Example: Samples from different sources.

consilience: A convergence of evidence. Consilience is the point at which multiple lines of evidence converge on a single conclusion. Example: The results of three different experiments lead toward one conclusion.

correlation: A statistical relationship between variables such that a change in one results in a change in the other(s). Example: The effect of cooling time correlates with metal tensile strength.

corroboration: Support or validation of a hypothesis.

deduction: A type of reasoning that goes from generalities to specifics to form a hypothesis based on what is already known. Example: Oil under a machine means the machine has a leak. There is oil under the machine. The machine has a leak.

dependent variable: *See* treatment variable.

empirical: Derived from experimentation or verifiable by observation.

experiment: Experiments, also known as trials, are an empirical method used to support or refute a hypothesis by testing under controlled conditions.

experimental run: *See* treatment.

factor: A process condition that affects an output. The independent variable and noise are factors with an influence on the treatment variable. Example: Settings on a machine, temperature, material, quantity, and mixture are factors.

hypothesis: An educated guess at the solution to a problem or the cause of an effect. A hypothesis should be falsifiable and have the ability to predict experimental results.

induction: A type of reasoning that uses observations to form hypotheses. It goes from specifics to generalities. Example: All machines found to have oil under them had a leak; therefore, a machine found with oil under it has a leak.

iterative: Involving repetition or repeating.

iterative inductive-deductive process: An iterative process alternating between repetitions of inductive and deductive reasoning. Deduction is used to form a hypothesis that is compared to empirical data; the results are then used as a basis for a new hypothesis using induction.

level: The value or setting used in an experiment. Example: Setting the factor temperature at 55°C and 65°C is an example of two levels.

lines of evidence: Following multiple lines of empirical evidence to achieve consilience.

noise: A source of variation. Noise may be known or unknown.

objectivity: Being objective; without influences due to personal prejudices or emotional influences.

operational definition: A clear process-oriented description of empirical terms that can be easily understood by others; often phrased in terms of testing or measuring. Example: The operational definition for *missed operation* is a production step that is described in the work instruction for a workstation but was not performed before the part was moved to the next workstation.

proximate cause: The cause of a failure or occurrence that is close to the event or nonevent that resulted in the failure or occurrence. There may be an underlying ultimate cause.

randomization: Performing experiments in a random order to minimize the effects of noise.

repetition: Measuring the experimental results more than once to account for measurement system variation. Example: The same part is measured five times with a micrometer to determine how much of the result is caused by variation in the calipers.

replication: The complete repeat of an experiment using the same experimental conditions as the previous experiment. Replication increases the precision of the results. Randomization is not used. Example: A quality engineer repeats an entire experiment in the same order and using the same conditions as the first experiment.

response variable: The result or output of an experiment. The response variable is represented by the x in the formula $y = f(x) + \xi$ and is also known as the dependent variable. It is the result of the manipulation of the treatment variable(s).

root cause: The underlying cause of an occurrence or condition.

root cause analysis (RCA): The investigation performed to determine a root cause.

scientific method: An empirical method of investigation using hypotheses and testing or experimentation to support or reject hypotheses.

treatment: The specific sets of conditions during an experiment. Also called an experimental run.

treatment variable: The variable in an experiment that is manipulated or controlled by an experimenter. The treatment variable is represented by the y in the formula $y = f(x) + \xi$.

ultimate cause: The underlying cause of a failure or occurrence.

$y = f(x) + \xi$: The mathematical formula for the relationship between the treatment variable x and the independent variable y. The f means x is a function of y, and the symbol ξ represents error.

Index

I

J

K

L

Q